P9-AOS-329
W9-CEK-987

Very Special Relativity

An illustrated guide

What I value in life is quality rather than quantity, just as in Nature the overall principles represent a higher reality than does the single object.

Albert Einstein

VERY SPECIAL RELATIVITY

An illustrated guide

Sander Bais

Harvard University Press 2007
Cambridge, Massachusetts
Londen, England

This publication is printed on woodstock Camoscio by Fedrigoni
Printed in Italy

First published in the Netherlands by Amsterdam University Press, Amsterdam

Design and lay-out: Gijs Mathijs Ontwerpers, Amsterdam

ISBN-13 978 0 674 02611 7
ISBN-10 0 674 02611 X

Library of Congress Cataloging-in-Publication Data
Bais, Sander.
Very special relativity: an illustrated guide / Sander Bais.
p. cm.
Includes bibliographical references.
ISBN-13: 978-0-674-01967-6 (alk. paper)
ISBN-10: 0-674-01967-9 (alk. paper)
1. Special relativity (Physics)--Miscellanea.
2. Special relativity (Physics)--Pictorial works. I. Title.
QC173.65.B35 2007
530.11--dc22
2007009144

Edition for the Netherlands: Amsterdam University Press,
ISBN 978 90 5356 965 8

Acknowledgements

I would like to thank the Yukawa Institute for Theoretical Physics of Kyoto University in Japan, where part of the work was done, for the inspiring hospitality; Jan Bais, Bernd Schroers and Joost Slingerland for thoughtful comments and suggestions; and Gerard 't Hooft for his preface. I thank Gijs Klunder for the inspiring layout and Laura van der Noort for her assistance. I am also grateful to the people at AUP for their guidance and patience during the realization of this book. Last but not least there are the students who over the years bombarded me with their wonder, disbelief and their almost irresistible logic, and who taught me how wonderful special relativity is.

Sander Bais

Preface

by Gerard 't Hooft

No intellectual hero has been more inspiring to our imagination than Albert Einstein when he discovered Special and General Relativity. Not only physicists are charmed by these beautiful constructions of the human mind, but also young students and the public in general. The ingenuity of his ideas became proverbial, even in Einstein's own time. So much so that even he himself once quipped: "I'm no Einstein...!" One consequence of this phenomenon is that we physicists nowadays receive numerous letters from people who think that they can outsmart Einstein, by "improving on" or "disproving" his theories.

Science, however, progresses in a different way. We don't really replace theories, we expand on them. A key element of science is that we also simplify. What looked complicated once is simple and straightforward now, and this is also what happened to Special Relativity. In a sense, it is just geometry. In comparison, the Euclidean geometry of triangles, spheres and cones is simple enough to visualize, so much so that it can be taught at high school. Special Relativity just happens to be the geometry of space and time. Simply add clocks to Euclidean geometry, and that's it! Well, not quite – there is something funny about light rays that makes spacetime geometry counterintuitive, and to visualize this in our minds requires a little more practice.

Many popular treatises on Special Relativity use words rather than diagrams or formulas. One would think that that would make things easier: People who are not accustomed to mathematics and geometry should find it easier to understand texts rather than equations. But that may not be true. When one hides the diagrams and the equations, talking about relativity becomes harder. So Sander Bais had the excellent idea to address the non-experts, the public, the young students, while making good use of geometric diagrams. The result was this marvelous little book. Once you understand how to read the diagrams, all of special relativity becomes beautifully clear. You can see at one glance that such a theory needs no "improvement" or "counter arguments": It is as useful as Euclidean geometry was to the ancient Greeks, and they both still are, up to this day. Special relativity in pictures. If you once found triangles, spheres and cubes fun to play with, you will certainly appreciate what you find here.

To my children, my father, and Vera

Table of contents

Introduction

No! Not another book on special relativity! Is my clock running slow? Is there any need for this book, so soon after the Centennial of the "Miracle Year" 1905, when the young Einstein brought all of physics into great disarray?

The saying goes that a smart person may know all the right answers, but the wise person stands out because he or she knows how to ask the right questions. Special relativity is such a delight exactly because the problem was basically to ask the right questions. You are heading for the contents of two of the most influential papers in all of physics, papers that shattered the classical notions of space, time, mass and energy: 'On the Electrodynamics of Moving Bodies', and 'Does the Inertia of a Body Depend on Its Energy Content?' Right from its inception it has been a challenge to express relativity in an ever simpler language, and that is what I try to do here.

This book is a "hands on" user manual aimed at an audience curious enough to want to know how relativity really works, with some knowledge of basic science and elementary highschool mathematics (in particular geometry). What is required more than anything else to enjoy this book is a playful mind. I have tried to make the content of the theory more accessible by presenting it in a "very special" way, using an easy-to-follow sequence of *spacetime diagrams*. I chose this particular geometric approach because images often speak for themselves and persist in memory, while algebra can be dense and easily forgotten. After all, "C'est le ton qui fait la musique."

We start our journey by explaining some of the basic principles of the theory, such as the notions of an event, a frame of reference, an inertial observer and a world line. Soon we come to the Einstein postulates. Subsequently we shall tackle a few well known paradoxes and their resolutions, encountering along the way simultaneity, causality, time dilation and space contraction, and the simple fact that there is such a thing as a universal maximal velocity. These examples highlight the counterintuitive nature of relativity. After a geometric intermezzo we move on to the notions of momentum and energy, culminating in the magnificent formula $E = mc^2$, which expresses the deep insight that mass and energy are equivalent. We even go beyond special relativity when we study the world of an accelerated observer who experiences a horizon. There we'll catch a first glimpse of the general theory of relativity, which Einstein completed about ten years after special relativity. We conclude with an epilogue situating Einstein's achievements in the greater context of physics as a whole.

Now let's get to work. I hope you will have as much fun pursuing these chapters as I did putting them together. To encourage you I have included a quotation of Einstein at the beginning of each chapter.

Sander Bais
Amsterdam, 2007

1 Basic principles

*It is a miracle that curiosity survives formal education.**

Space + Time = Spacetime

We all agree that space and time are with us, that we are moving about in space and time: They constitute the arena in which our lives unfold. Nevertheless they are untouchable and we perceive them only indirectly through our senses, which make us aware of the things that are happening. Seeing objects at different distances gives us a spatial awareness, while observing change creates our notion of time. And as stars, golf balls and dogs move continuously, our perception is that space and time are continuous. We are not living in a stroboscopic disco-like reality.

In many ways the notions of space and time are fundamentally different. We cannot go back to the past and interfere with it, while the future is equally inaccessible. Our active being is confined to the fragile interface between the two, the present. In space we can only be in one place at a given time too (though many try to defy this basic law), yet we can choose to move from one place to another or to stay in a specific spot. Time is measured with a clock while space is measured with a meter stick, two entirely different kinds of instruments.

These facts do not keep us from representing the notions of space and time in a simplifying picture, a map with space and time coordinates on it, shown in the figure. In the experts' jargon this is called a Minkowski diagram; it is a kind of conceptual map, not of the world, but of what is going on in the world.
This book is about what these spacetime maps mean and what they look like from the points of view of different observers.

* At the beginning of each chapter you will find a quoted text by Einstein.

time

0

space

Events

We have only drawn a small piece of space and time; you should think of space and time as extending out to infinity in both directions of the graph plane. We have chosen to put time along the vertical axis and space along the horizontal. This means that all of three-dimensional space – height, depth and width – has been reduced to a single space axis, implying that we can only discuss forward and backward motion in one spatial dimension, very much like a train on its track. Yet such a drastic amputation of space will not really affect our ability to convey the essentials of special relativity.

What can we do with a spacetime diagram? What do points, lines, curves and domains mean? Let's start with the simplest ingredient, one point in spacetime. What does it correspond to? A point defines a particular place at a particular instant in time: It represents an *event*. You clapped your hands precisely there and then! You dropped something, you fired a gun, or you ran into somebody. Our spacetime world is densely populated with events, and these correspond with points in our diagram. Conversely one could say that spacetime is the collection of all possible events. We see events as connected in time: We perceive the motion of a tennis ball not as a set of distinct events, but as a continuous sequence of events, called its motion. Things that move correspond to paths or curves in our spacetime diagram. Similar pictures can be used to represent the time development of virtually anything. If that thing is the profit of a company, then time would run horizontally and the units along the vertical axis would typically be a million dollars, with the negative part of the axis taking care of the losses. To visualize the development of the population of various countries, we can draw curves with the number of people along the vertical axis. But before going into those curves, let me dedicate a few words to the scale of things.

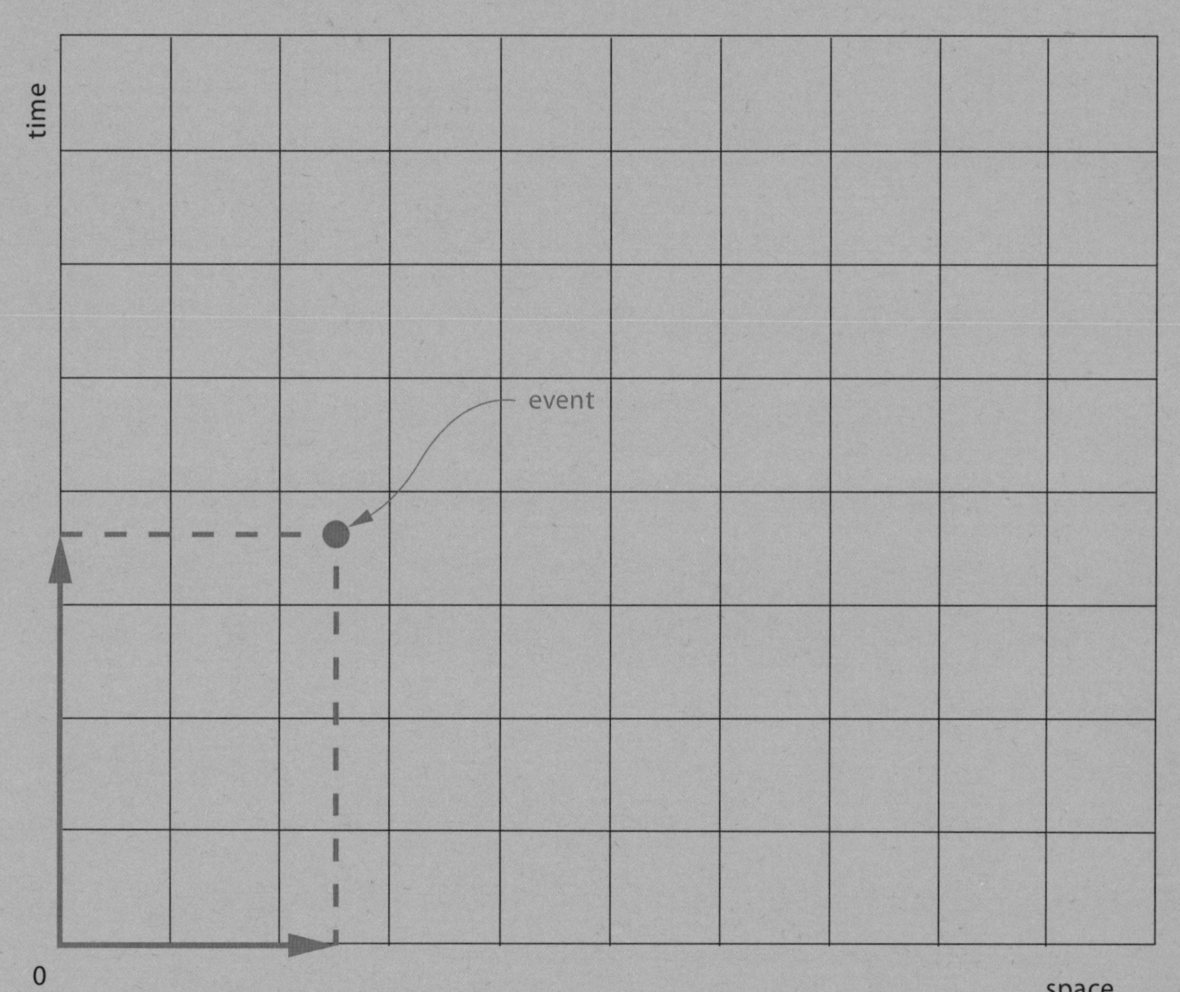
time
event
0
space

Setting the scale

We have drawn a grid of horizontal and vertical lines in spacetime. The grid provides us with a coordinate system, a frame of reference, which allows us to conveniently label individual events indicating where and when they occurred. It is just like the grid on a city map, which allows you to localize yourself in space. You may also think of the way chess players talk: They use coordinates – e.g. "e2 to e4" – to describe their moves on the chessboard.

The mesh has a certain scale or unit for each of the axes. On a city map, steps of half a mile will typically label both axes. On a map of countries the unit is perhaps a hundred miles or so. Now we want to set the scales along our time and space axes, and we should choose a convenient measure so that the things that are relevant to us become clearly visible and distinct. The phenomena we are about to discuss are related to the relative proportion of distance and time – in other words, to the notion of distance per time, which is by definition velocity. So the relative scale between the time and space axis is set by the scale of the velocities that are relevant in the context of special relativity.

We are about to learn that this is not our typical everyday human velocity scale, say of meters per second or miles per hour. Not the speed of a car, of an airplane, or the speed of sound – no, it is a very unique and as we will see even universal velocity we will pick: the *velocity of light*, which is conventionally denoted by the letter c.

What is so special about the velocity of light? Well, at first nobody realized there was anything special about it, until Einstein recognized it as a universal constant of nature. Before that, we only knew that the velocity of light was finite, just like any other velocity. Around 1850 its value was determined in a clever but simple experiment by the French physicist Fizeau (see box on page 19). He found the velocity to be very close to 300,000 kilometers per second. Since October 21, 1983, c is actually *exactly*

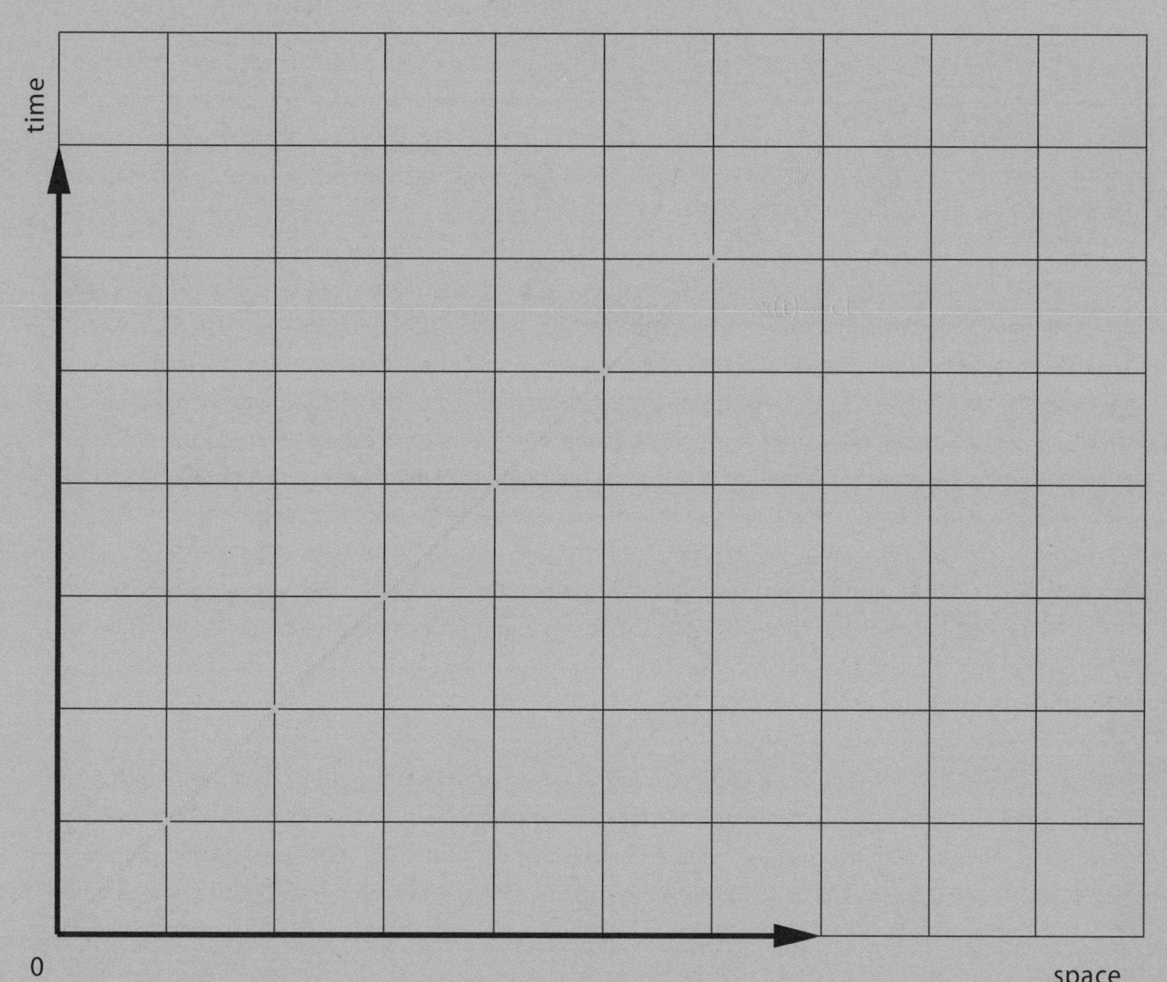
time
0
space

299,792,458 meters per second, because c is now used to define the meter. This is a huge number indeed, which explains why we perceive it as basically infinite: if we turn on the light, it appears to fill the room "instantaneously". But that is an illusion, because the light has to propagate from the light bulb to the walls and that takes some time, though it takes less than a millionth of a second. In most everyday circumstances, "instantaneous" is therefore not such a bad approximation.

To set the scale of our spacetime map we do the following. Suppose we use the second as unit along the time axis, then we put the distance light travels in one second as unit along the space axis. Now, if we send a very short pulse of light (or even better: a photon, a single quantum of light) in the space- or x-direction, it traces a path on our spacetime map that corresponds to the yellow arrow indicated in the figure. We can write this as $x(t)= ct$, which can be translated into proper English as: the position x at time t equals c times t. If $t = 1$ (sec) then $x = c$ km, if $t = 4.5$ then $x = 4.5$ times c km, et cetera. Because the combination $w = ct$ will occur quite often, *we will from now on define w as the time coordinate*, and use w instead of t. Note that if something moves with a constant velocity, it traces out a straight line in the diagram, because in twice the time it travels twice as far. And it is the slope of the line that determines the precise velocity, as we will see next.

Brainteasers: 1. What would a pulse of light which lasts one second look like in the diagram?
2. Draw the line of a photon wich moves to the left.

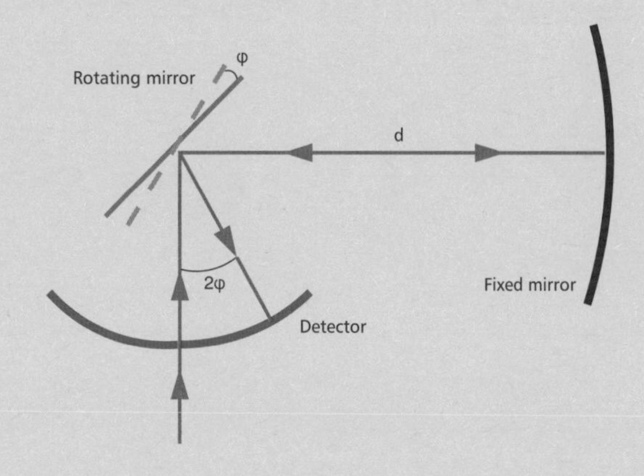

Measuring the speed of light

The first proposals for determining the velocity of light go back to the Dutchman Isaac Beeckman in 1629. Ole Rømer made the first quantitative determination using astronomical observations in 1676. The French physicists Fizeau and later Foucault performed the first earthbound experiments around 1850. A schematic of a typical experiment is given in the figure. An incoming light ray is reflected off a mirror that rotates with a given angular velocity of ω degrees per second. After continuing over a fixed distance d (in the original experiment about 10 miles), the ray gets reflected by a fixed mirror. When after a time Δt the light returns to the rotating mirror, the latter will have rotated over an angle $\varphi = \omega\Delta t$ and therefore the reflected beam will have an angular deflection of 2φ, which is measured. The velocity of light is then simply given by $c = 2d/\Delta t = 2d\omega/\varphi$.

In 1886, before the advent of relativity, Michelson and Morley conducted another very important experiment concerning light, which showed that the velocity of light was independent of the direction in which the light beam traveled. This result implied the absence of an *ether* – a background cosmic fluid – completely consistent with a basic postulate of special relativity, as will be discussed later.

World lines

We alluded previously to the fact that objects trace out a continuous path or curve in spacetime. Now the word path has a strong connotation with a path through the woods or the city, a path through space. Therefore when we talk about a path through spacetime it is customary to call that path a *world line*. In the figure we have depicted various world lines. They all start at time zero at the point marked $x = 0$, which is referred to as the *origin*. (There is no deep meaning attached to this point: it is not that space and time originate there, but rather a quite arbitrarily chosen reference point in our coordinate system.) Of course, these world lines move forward in time, which is reflected by the fact that they never bend down.

First look at the black arrow which coincides with the time axis. It simply depicts somebody or something at rest, sitting at $x = 0$ and staying there forever, immobile. The yellow arrow is the familiar world line of a light pulse or photon. The other straight lines correspond to objects that move with other constant velocities – constant, because the distance traveled is always proportional to the time the object has been traveling. The red arrow could be somebody who travels at a velocity v which equals $v = 3/4\ c$, because at any given time he has traveled 3/4th of the distance that the light pulse has covered during that same time. This is particularly clear at time $t = 4$, where the red traveler has moved three units of distance while the light pulse has traveled four. By the same reasoning one concludes that the green traveler (called "the phantom") must be traveling at twice the velocity of light. Finally there is that wiggly blue world line. It describes a traveler who moves back and forth with varying velocity: she is speeding up and slowing down, as you can see. At each given instant she has a velocity which is determined by the slope of the tangent to her curve at that very instant. So the world line provides an accurate account of the history of the motion of a traveler.

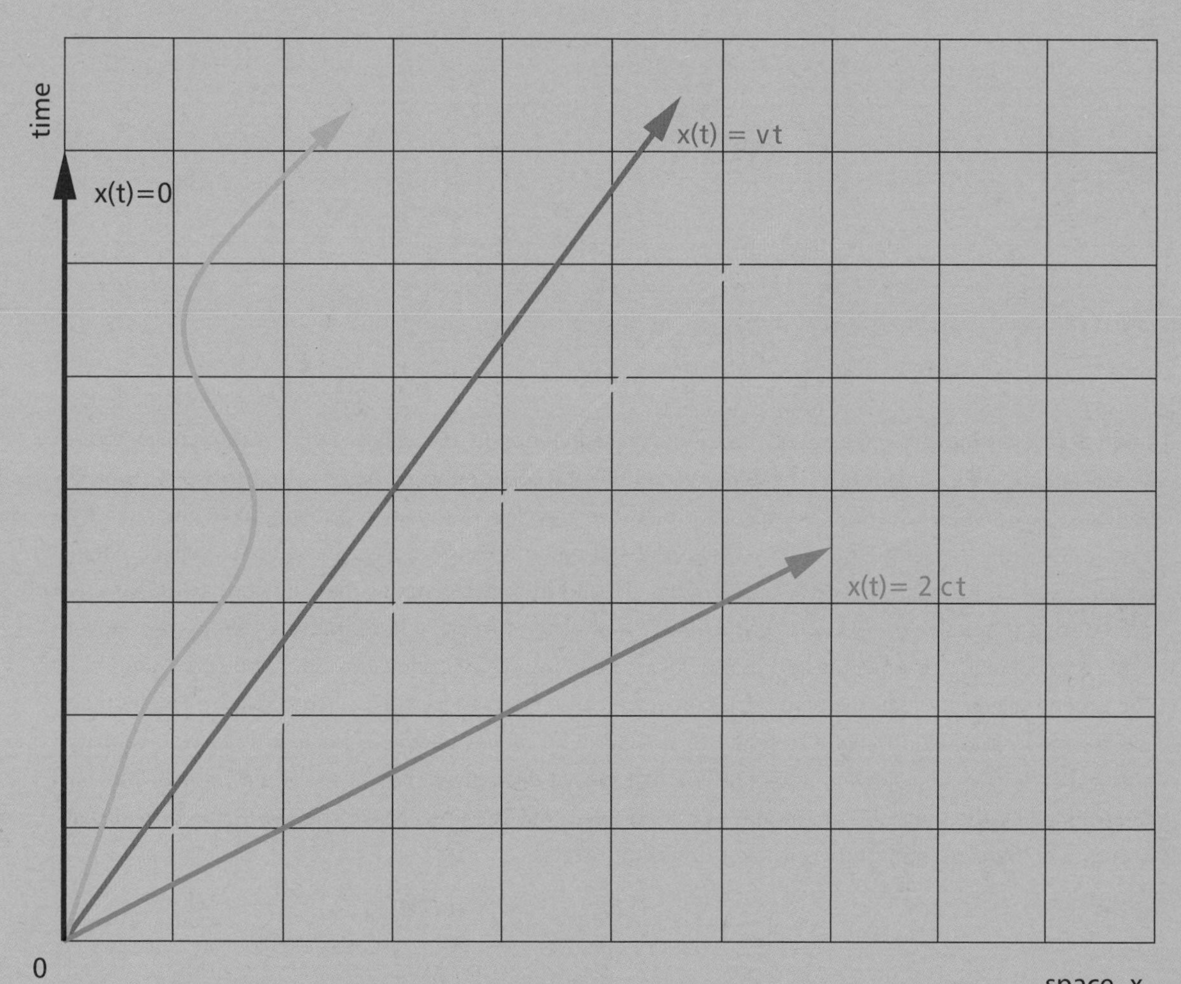
time
x(t)=0
x(t) = vt
x(t)= 2 ct
0
space x

The postulates

What we really want to know is where Einstein took us. Therefore I am going to present you with a straight statement of the facts, not a long-winded story of how it came about and what profound debates the scientists of that era had to go through before they could handle the theory and convince themselves of its profound meaning. This book is not a biography; I merely want to convey the essentials in a "do it yourself" fashion. In our presentation we will rather dogmatically stick to the language of spacetime diagrams. This rendering should allow you to deal with some of the questions that certainly will arise along your world line and to answer them yourself by tinkering with the diagrams.

We will take a minimal formulation as our starting point, and from there extend our understanding of what it all means and why relativity is so special and shocking at the same time. This basically means that we are going to start where Einstein ended his story of special relativity, when he summarized the theory very efficiently in two postulates, two fundamental assumptions about nature.

The first postulate concerns two frames of reference or two (collections of) observers traveling with a *constant speed* with respect to each other. Such frames are called *inertial frames*. That's what the 'special' means: no accelerations, only constant relative speeds. The postulate then says that if those different observers each do experiments in their own frame of reference, they will discover the same laws of physics (if they are smart enough); they will arrive at the same equations describing the laws of motion, gravity, electromagnetism, and the other forces. That does not sound too alarming, does it? It sounds perfectly reasonable, and indeed Einstein was not even the first to make such a statement. Galileo Galilei made a similar observation some three hundred years earlier, when considering "fish and ships" in his *Dialogue concerning the two chief world systems*:

For observers moving with constant velocity with respect to each other, the following postulates hold:

1 The laws of physics are the same.

2 The velocity of light in vacuo is the same.

> "Shut yourself up… in the main cabin below decks on some large ship... Take a large bowl of water with some fish in it… While the ship is motionless… the fish swim indifferently in all directions… When you have observed all these things carefully…, have the ship proceed with any speed you like, so long as the motion is uniform and not fluctuating this way and that. You will discover not the least change…, nor could you tell… whether the ship was moving or standing still."

Later we will see that Einstein's postulate is far from straightforward if we critically analyze the similarities and differences between the Newtonian laws of mechanics and the laws of electromagnetism as expressed by the Maxwell equations, focusing on what they look like in different frames of reference.

The second postulate states that the velocity of light in vacuum (i.e. in "empty space" – not in some medium where there may be all kinds of complicated interactions going on), is the same for every observer, irrespective of with what (constant) velocity he or she moves. Now that *is* strange, if you think about it. It goes right against our intuition about velocities, and for that matter right against Newton's theory. If I am riding a bike at 10 miles an hour, and I throw a candy bar forward to my wife with a velocity of 15 miles an hour, then my wife, who is standing on the sidewalk, will catch the bar and say that she received it with a velocity of 10 + 15 = 25 miles an hour. We are happy to agree, because that's just the way it is. Excuse me, let's be precise: that's the way it *was*…

What Einstein tells us is that if I am riding a very fast train that is traveling at half the speed of light, $v = \frac{1}{2}c$, and with my laser flashlight I send a short pulse to my partner at a station far away, sure enough the pulse is moving with the velocity of light with respect to me. Therefore by our previous intuitive reasoning we would expect my partner on the platform at the station, if she were to measure the velocity of the pulse, to find the answer $u = c + \frac{1}{2}c = 1\frac{1}{2}c$. But now Einstein comes along and bluntly says: No! She also measures $u = c$. That is strange indeed, strongly counter to our intuition to say the least.

How can this be? How can such a simple argument fail to be correct? That's how most of the physicists at the time also reacted. If Einstein is right, then the price we have to pay will be high – and that is indeed the way it was. You see, velocity is distance (space) divided by time, and to implement the equality of the speed of light for all observers we have to go deep in a conceptual sense and reconsider our notions of space and time at the most basic level. This is what we'll want to come to terms with. It is hard to defeat prejudices, and we will have to work through some diagrams to rid ourselves of some very persistent but wrong intuitions.

2 The relativity of simultaneity

The whole of science is nothing more than a refinement of everyday thinking.

Frames of reference

First we want to know how different observers who are at rest with respect to each other, set up a frame of reference or coordinate system. In fact a frame of reference corresponds to a large number of "observers" who are at rest with respect to each other, such as the passengers sitting in a moving train or the collection of people standing on the platform. They all have clocks and meter sticks and are so kind as to do measurements if we ask them to do so, and willing to very obediently report their findings to us: They are perfect subordinates.

We start with two observers who are given identical clocks and rulers. They are indicated in the figure by two black arrows: apparently they sit still and are some (large) distance apart. They want to calibrate their clocks so that they can share their time measurements in a sensible way. How should they go about it? That is depicted in the next figure.

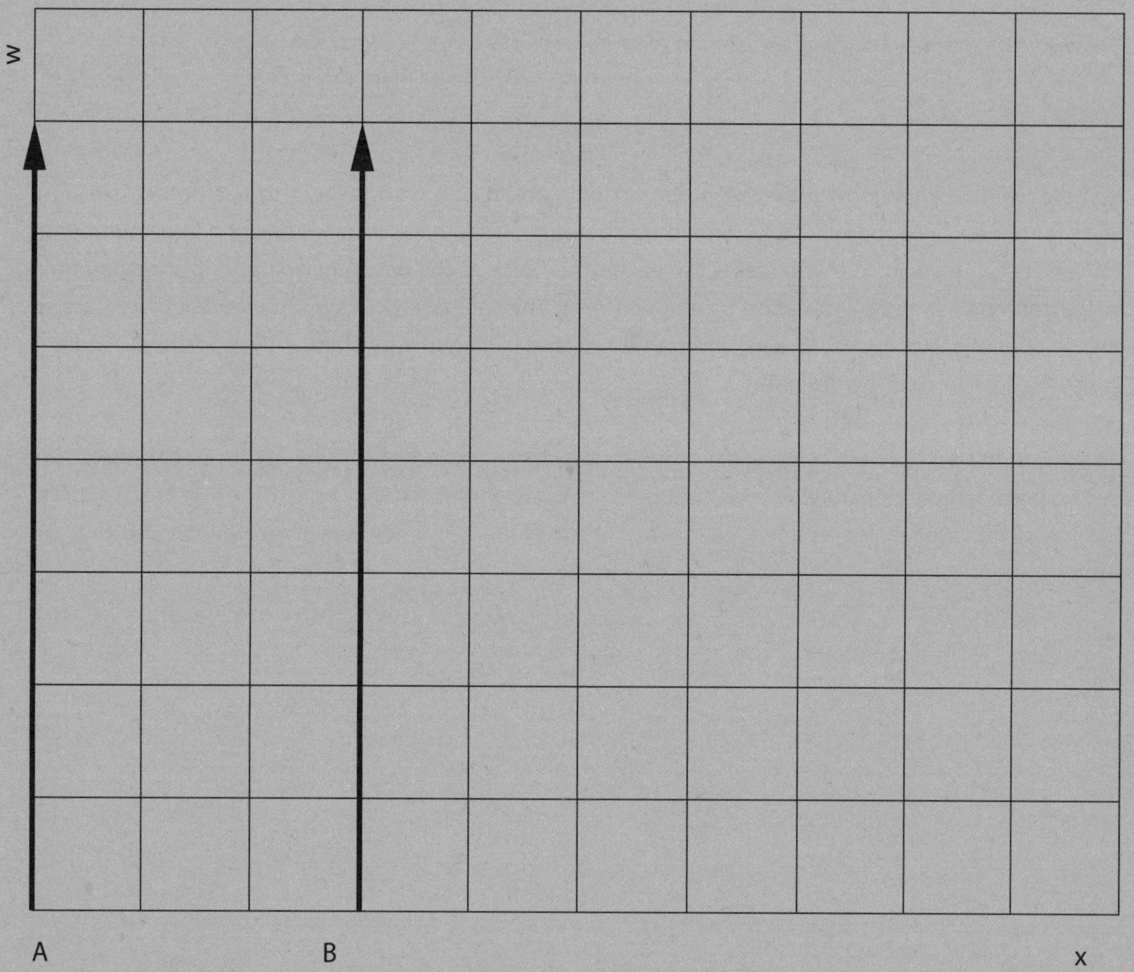

w

A

B

x

Calibration of clocks

It is best to think about the calibration as a real physical experiment. In later chapters we will often encounter "thought experiments", which make complete sense theoretically but would be hard to perform in real life. In this case Einstein gave a simple recipe that makes complete sense: at time $w_A = 0$, observer Apollo sends a light signal to observer Bacchus, who clocks it at time $w_B = w_1$ and with a mirror reflects the signal back to Apollo, who clocks its arrival at time $w_A = w_2$.

Now the time which is simultaneous to w_1 for Apollo is the time instant halfway between the sending and arrival of the light signal, i.e. $w_1 = ½\, w_2$. This is not surprising, because it confirms what we already knew from the grid we had drawn. Furthermore, because we know the total time a signal takes to travel forth and back is proportional to the distance between the two observers, this approach can be used by a larger set of observers to set up the whole grid this way.

Note, however, that the notion of simultaneity for a group of observers who are at rest with respect to each other does not imply that they would all observe simultaneous events at the same time! One has to take into account the difference in the time the signal takes to travel from the event to the different observers. This means that different synchronized observers can take down pretty dissimilar times, yet after adjustment for the signal's traveling time they will all ascribe the same moment in time to the event.

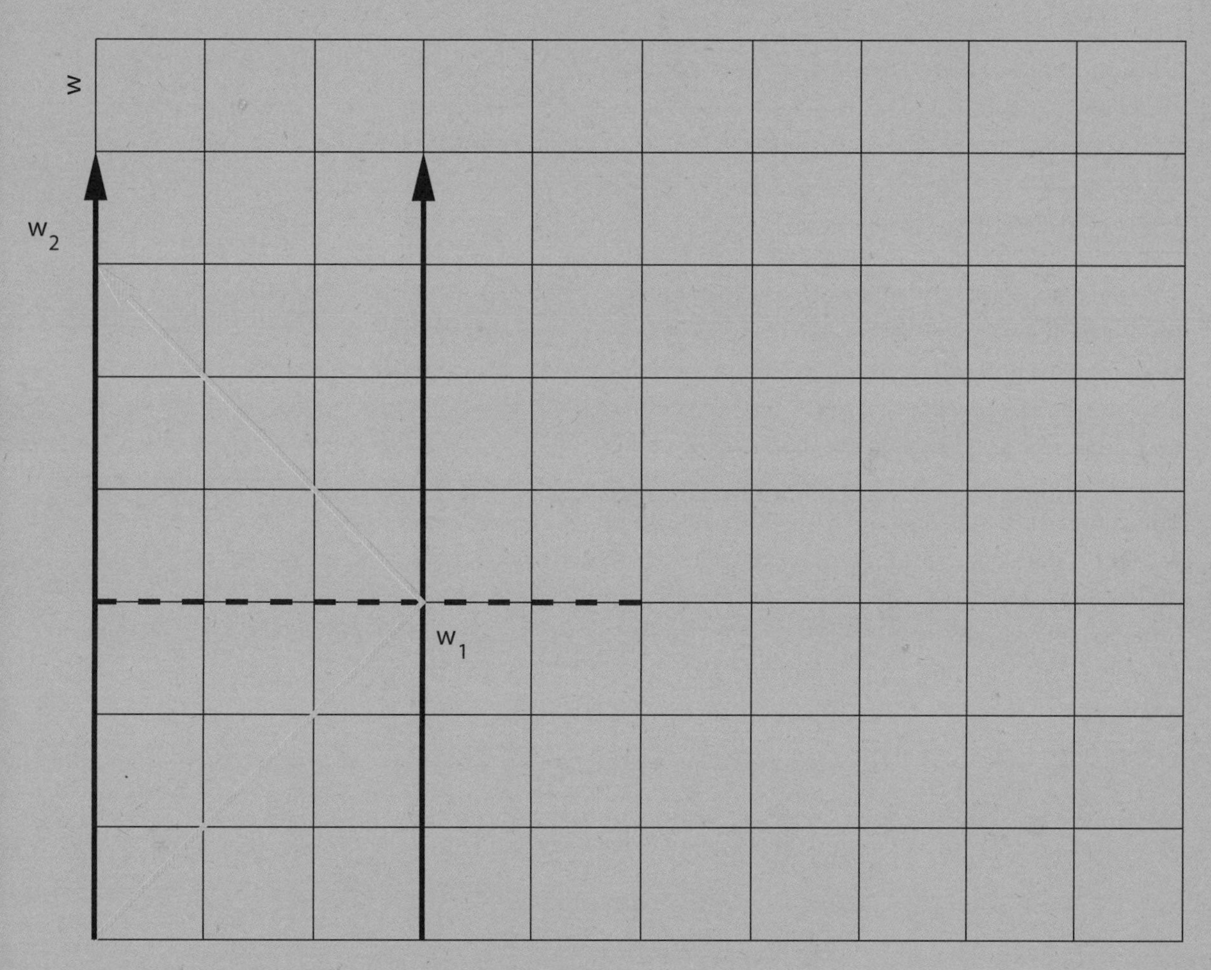
w
w2
w1
A
B
x

Moving frames

Now that we have set up one frame of reference, we will use the same recipe to set up another inertial frame that belongs to a set of observers who all move with the same (non-zero) velocity. Accelerated frames or rotating frames are not inertial frames, because the velocity is not constant – even when a ball attached to a string moves in a circular orbit with fixed orbital velocity, that velocity is not constant because its direction changes continuously. In those cases the relativity postulate does not hold. This is why it is necessary that the red world lines of Arnold and Britney are straight lines. They too want to calibrate their clocks, to set up the red frame, accurately following Einstein's instructions.

So Arnold and Britney do the same experiment. When depicting this in a diagram we have to adhere to Einstein's second postulate, which says that the velocity of light is the same for all observers. This means that the world lines of the light emitted by the moving observers Arnold and Britney should appear in the spacetime diagram under the same angle (45 degrees) with the axes as in the rest frame. (Just like in the next figure.)

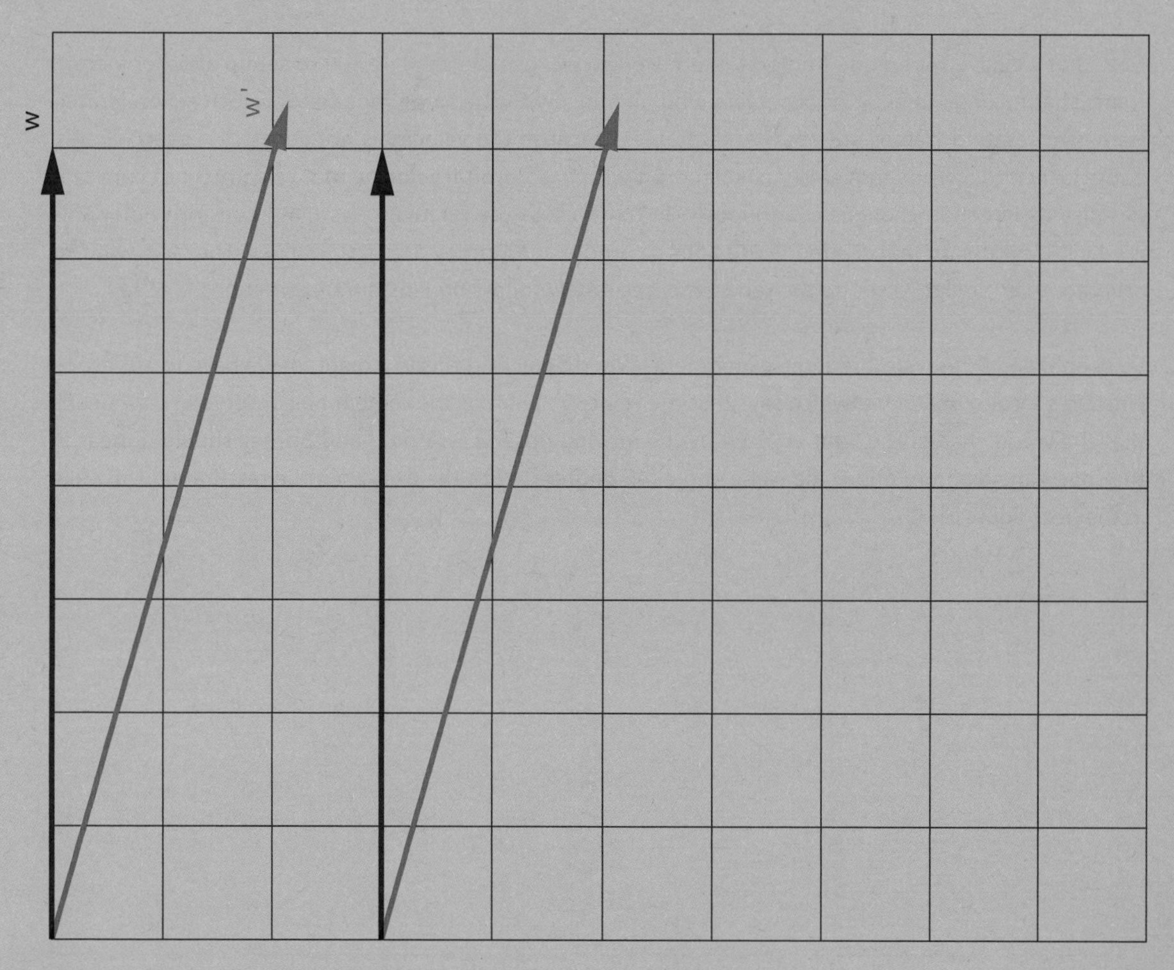
w
w'
A
B
x

The relativity of simultaneity

Arnold sends the light signal at time zero, Britney receives and reflects it at time $w'_B = w'_1$ and Arnold welcomes the signal back at $w'_A = w'_2$. To find out which time on Arnold's world line coincides with w'_1 on Britney's, we have to apply the same logic, leading to the instant halfway: $w'_A = ½\, w'_2$. This is nice: we apply the same procedure as for Apollo and Bacchus and thereby act in agreement with the "relativity" principle.

However, now something drastic has happened, which becomes evident if we look at the dotted red lines. These lines are per definition the lines of "equal time" in the red frame: they connect events that are simultaneous for the red observers. We could also say that these are the lines along which red people measure distances, and the dotted line through the origin is nothing but the new space or x'-axis, where the prime refers to the red frame. In some sense, the meaning of length presupposes the notion of simultaneity. If we want to measure the length of a table, we put a meter stick alongside it, and if we want to do the measurement properly we have to read the stick off at both ends of the table at the same time – otherwise the table (or the meter stick) could have moved between the reading off at one end and at the other, and the measurement would be meaningless.

The startling fact illustrated by the picture is that the space axes for the rest frame and the moving frame are not parallel, so things that are simultaneous (i.e. happen at the same time) in the black frame, connected by a horizontal black line, are generally not simultaneous in the red frame! A first important lesson to be learned from this is that the notion of simultaneity or "at equal time" is frame dependent: whether or not two events take place at the same time depends on which set of observers is measuring them. Simultaneity is a relative concept.

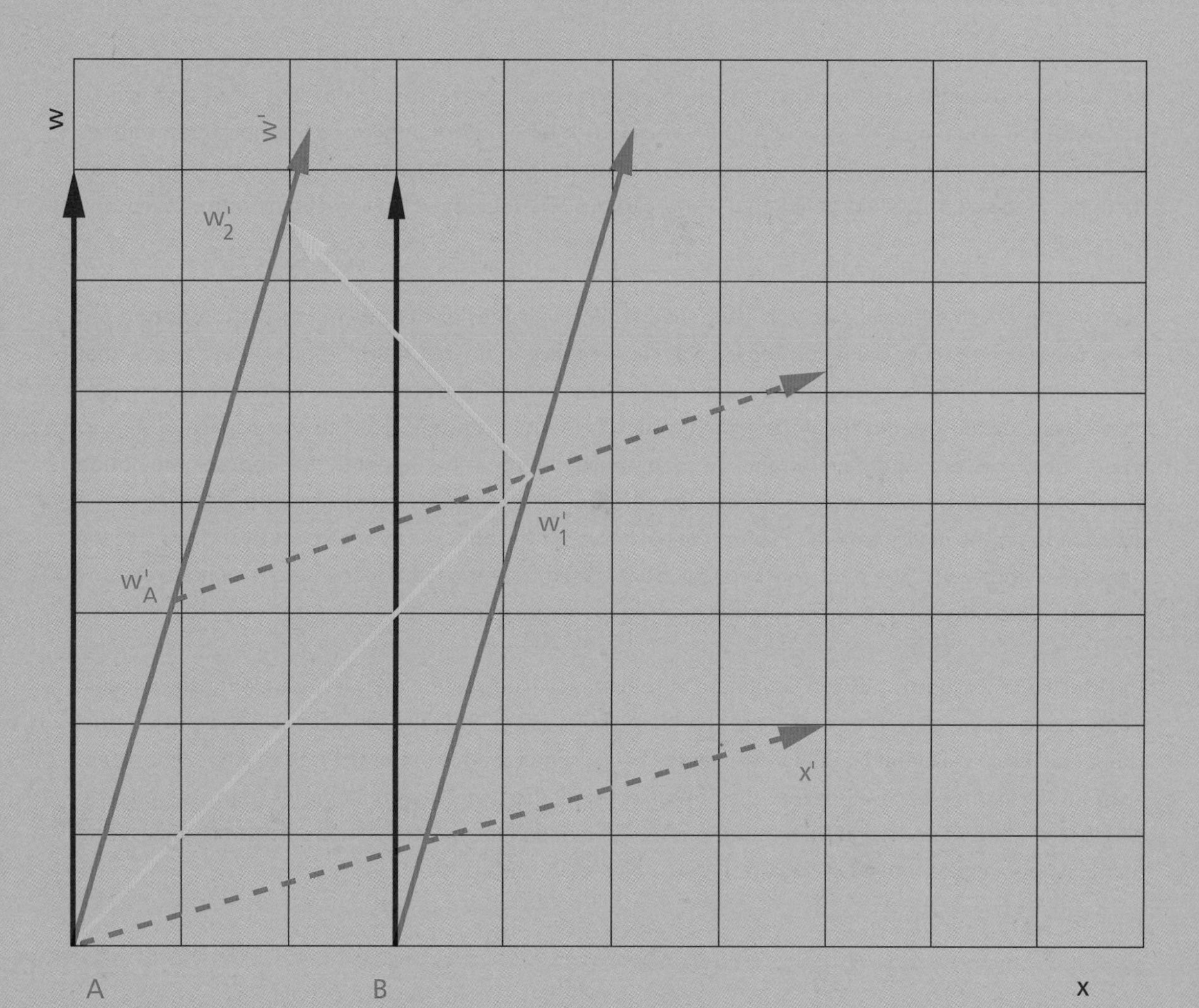
w
w'
w'2
w'1
w'A
x'
A
B
x

One spacetime, many inertial frames

We arrive at the following picture: Spacetime can be covered with all kinds of grids, but the grids of the inertial frames that move with respect to our black frame are oblique, like the red one in the figure. We see that the angles the two new (moving time and space) axes make with the old ones are equal, and so are their angles with the world line of a light signal. Therefore a point on this world line will again have equal components along the x'- and w'-axis. We are not yet worrying about what scale to put along these oblique axes.

Now we see what happened: with Einstein's second postulate, we have lost the absolute separation of space and time! Their relationship turns out to be dependent on the velocity one is moving at, so that it is better to refer to one entity that is the same for all observers: not space, not time, but spacetime.

It is gratifying to see that we can gain these rather surprising insights just by using qualitative arguments. Even so, at this point it is good to add a quantitative remark concerning the angles and slopes of the red grid. If the red travelers move with velocity v, the distance traveled during time t equals $x = vt$, and with the aforementioned $w = ct$ this becomes $x = vw/c$. This can also be written as $x/w = v/c$, which is by definition the tangent of the angle between the w-axis and the w'-axis. The *relative velocity parameter v/c is usually denoted by* β (beta), and it will be used extensively in the rest of this book. As a ratio of two velocities, β has no physical units: it is just a number. Note that also the slope of the angle between the x-axis and the x'-axis is equal to β.

Brainteaser: The figure suggests that the rest frame is very special. Draw another figure with black and red grids, where the red observers have the perpendicular grid.

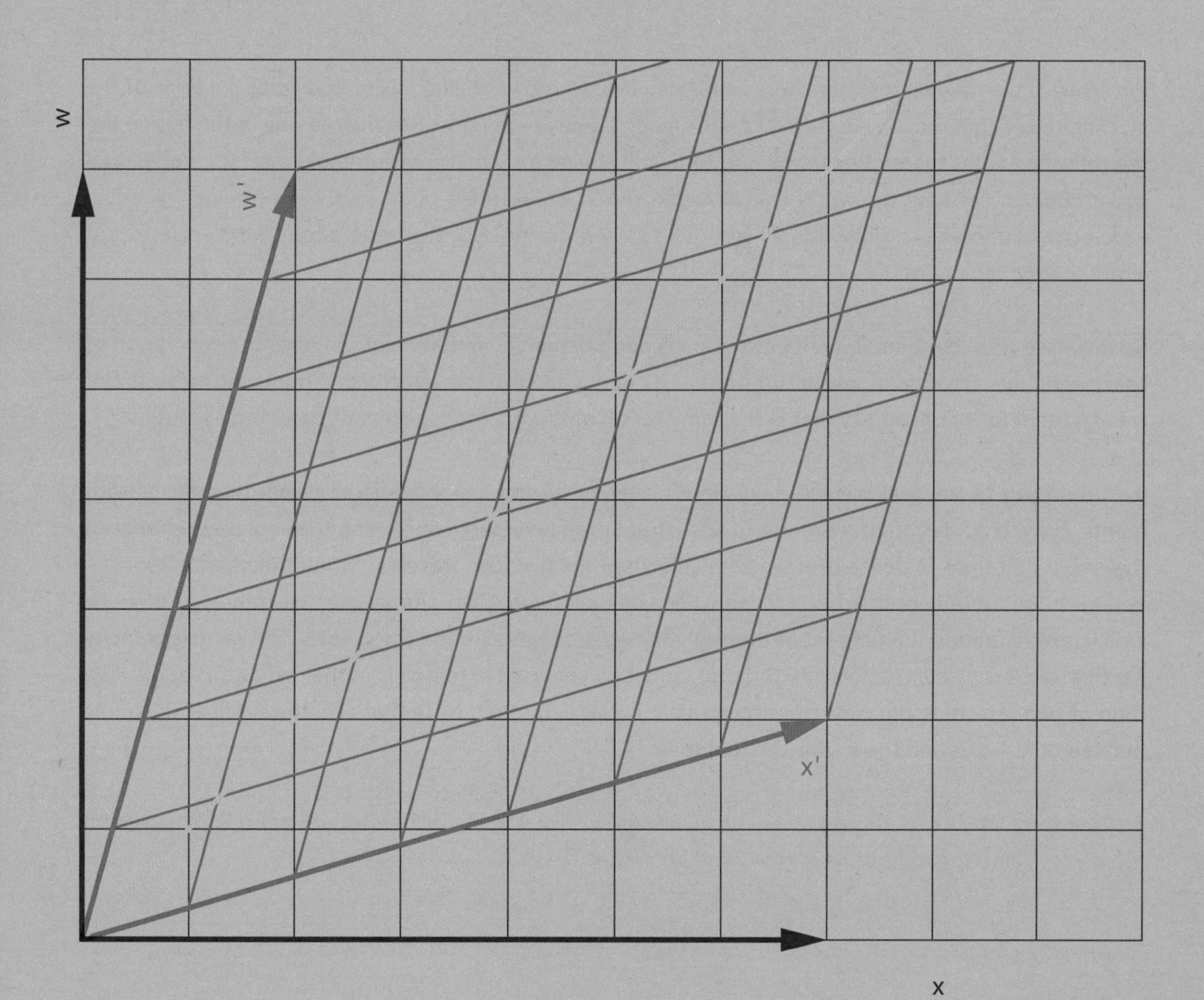
w
w'
x'
x

What's new?

This change in the very structure of space and time is so important that before discussing more of the consequences, we shall sit back for a moment for a comparison with the Newtonian theory, or should I say 'frame of mind'. To make a clear distinction I will consistently give these non-relativistic figures a grey background. We see that the rest frame looks the same, and in this frame the lines representing the red observer and the light pulse are also the same. But according to Newton a light pulse is nothing special: if the moving Arnold flashes his laser pointer, the signal moves with respect to him with the velocity of light c, but for the observer at rest that signal will move with velocity $c' = c + v$, represented by the second yellow/red arrow on the right. With Newton the velocity of light is not universal and its world line depends on who sent the signal. Now we see that the equal time lines are horizontal for all frames: *in Newton's theory it is time that is universal, and not the velocity of light.*

From this figure one may also infer how the coordinates (w,x) of an event in the rest frame are related to the coordinates (w',x') of that same event in the moving frame. We see that $w' = w$, and $x' = x - vt = x - (v/c)w$. This is the so-called Galilean transformation, which links the coordinates of two frames that move with velocity v with respect to each other. The word "transformation" indicates a sort of "translation" between the primed and unprimed quantities. Further on we will set out to find a similar relation between different frames in Einstein's theory.

Brainteaser: Use the same recipe as before to show that due to the non-uniqueness of the velocity of light, equal time lines for the red observers do become horizontal.

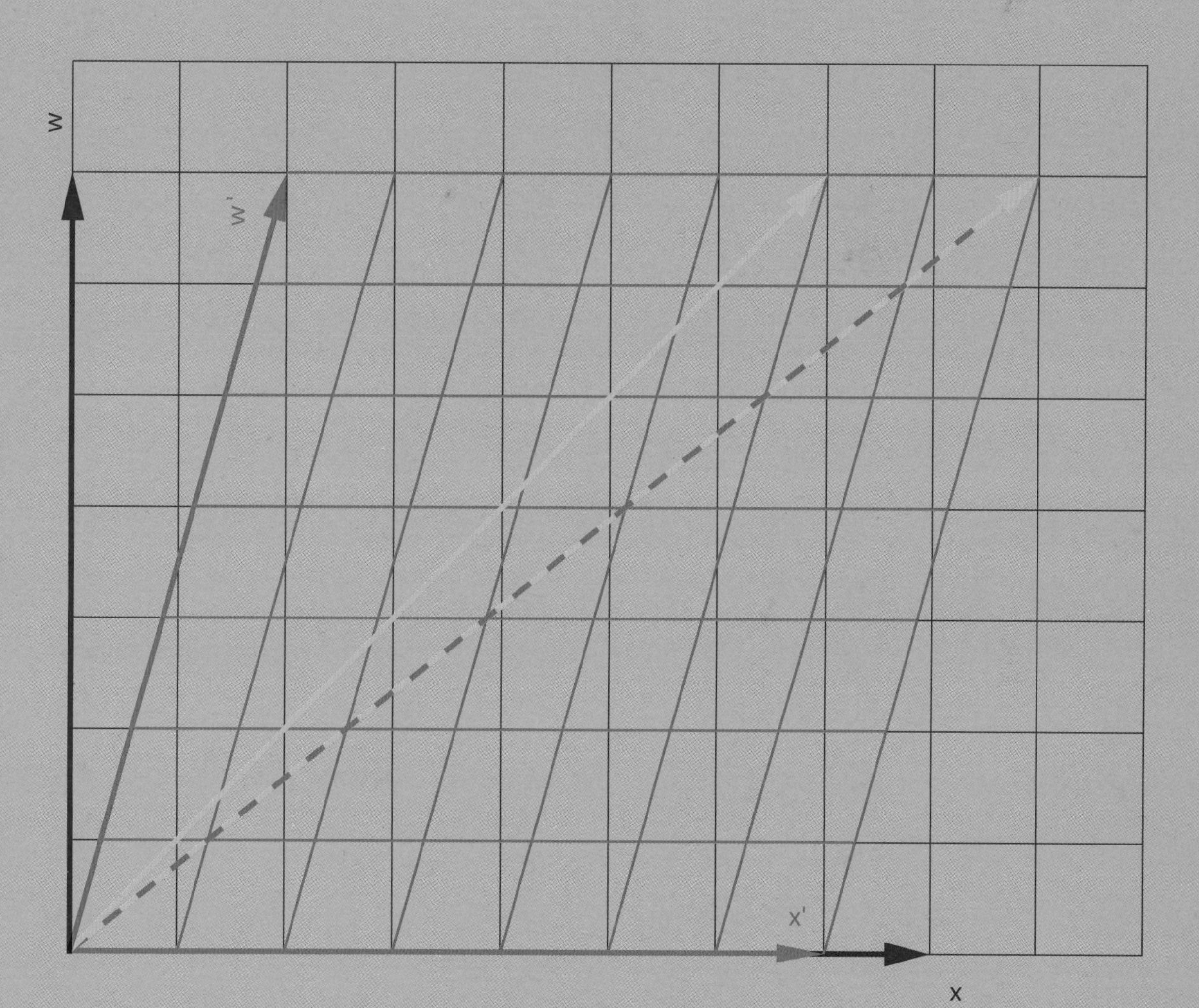
w
w'
x'
x

3 Causality

I never think of the future – it comes soon enough.

Causality lost?

Let us now consider two events, labeled 1 and 2 in the figure. Imagine 1 to be naughty Nigel carrying a gun entering a room, and 2 Auntie Augusta getting killed. From the point of view of the black frame there is nothing implausible about the hypothesis that naughty Nigel killed Auntie Augusta, because the diagram shows us that 1 happened after two time units and 2 after three. But now look at the sequence of events from the point of view of the red observers. First they see Auntie getting killed (after one time unit), and then they see Nigel entering the room (after two time units). The sequence of events is reversed. At first sight this appears to be a fatal inconsistency in the theory: how can the time ordering of events be relative? Doesn't that mean that Einstein went a bridge too far, and with his second postulate sacrificed the cherished notion of causality? Causality is not negotiable, because the whole of physics rests on it. We don't like to think of effects followed by causes. Not because of some narrow-minded scientific prejudice, but rather because it would lead to disastrous contradictions that violate any sense of reality. Imagine somebody firing a gun and killing somebody else. If we reverse the situation so we first see the person being killed and only at a later instant see the firing of the gun, we could in principle interfere in order to "prevent" the fatal shot, but the victim is already dead... This is nonsensical.

To appreciate what is going on in special relativity, where apparently we are faced with a demise of causality, we first must make a digression to gain another deep insight. It concerns the properties of velocities which follow from Einstein's postulates.

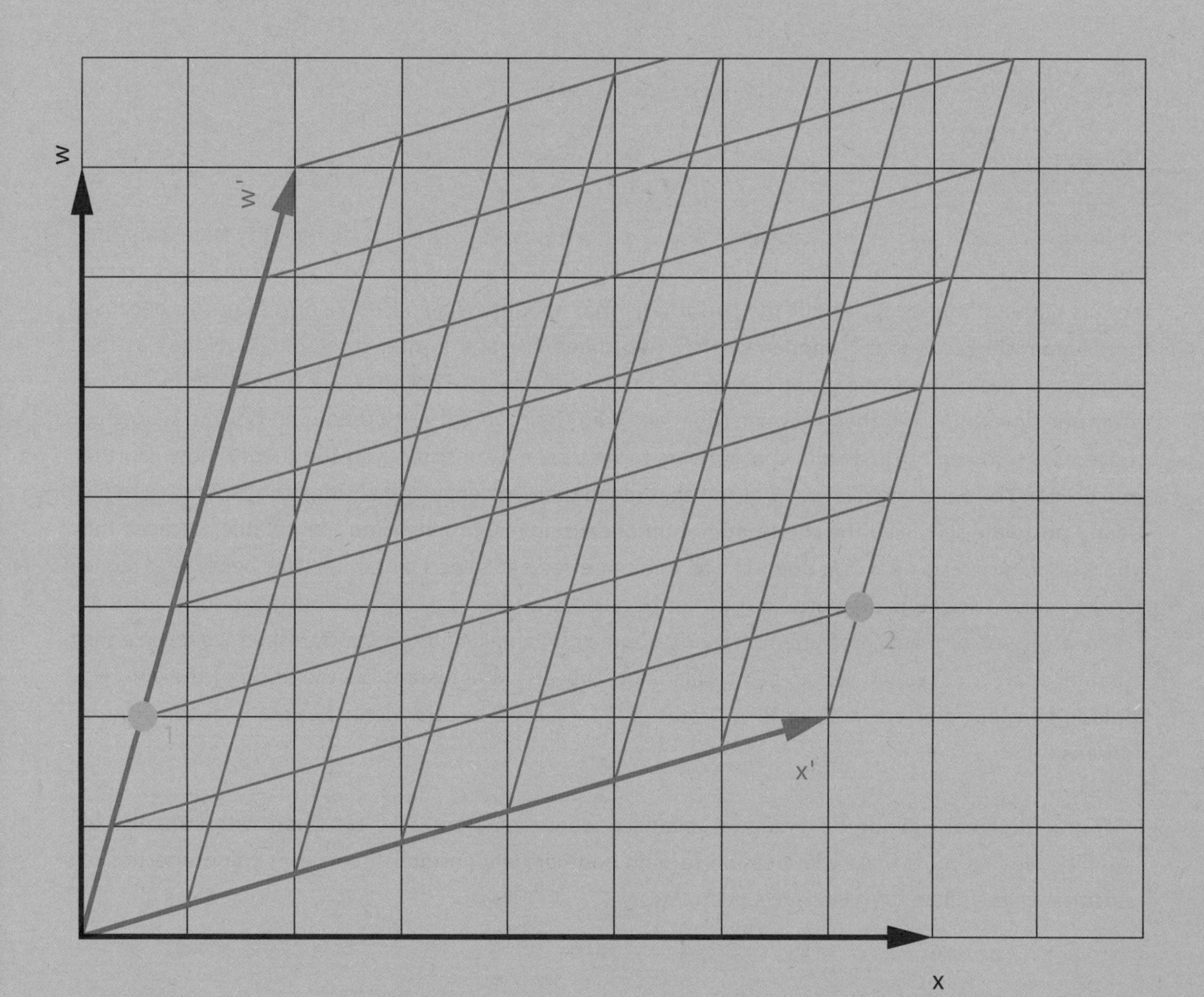
w
w'
1
2
x'
x

Adding velocities a la Newton

We start with describing the Newtonian perspective on the problem of how to add velocities from the point of view of different observers. We consider the following thought experiment. In the figure the red arrow describes a red (Super Shinkansen) bullet train moving with $2/7\ c$ – two-sevenths of the velocity of light. In the train a blue-eyed girl runs forward, also with $2/7\ c$ (i.e. two-sevenths of the speed represented by the yellow/red arrow on the right for observers within the red frame). The result is the blue arrow, and indeed in the black frame this arrow corresponds with a velocity of $2/7\ c + 2/7\ c = 4/7\ c$ (with respect to the yellow arrow, which represents the speed of light in the black frame). This is in perfect agreement with our naive (Newtonian) expectations.

Let us now turn to the same exercise with Einstein's point of view.

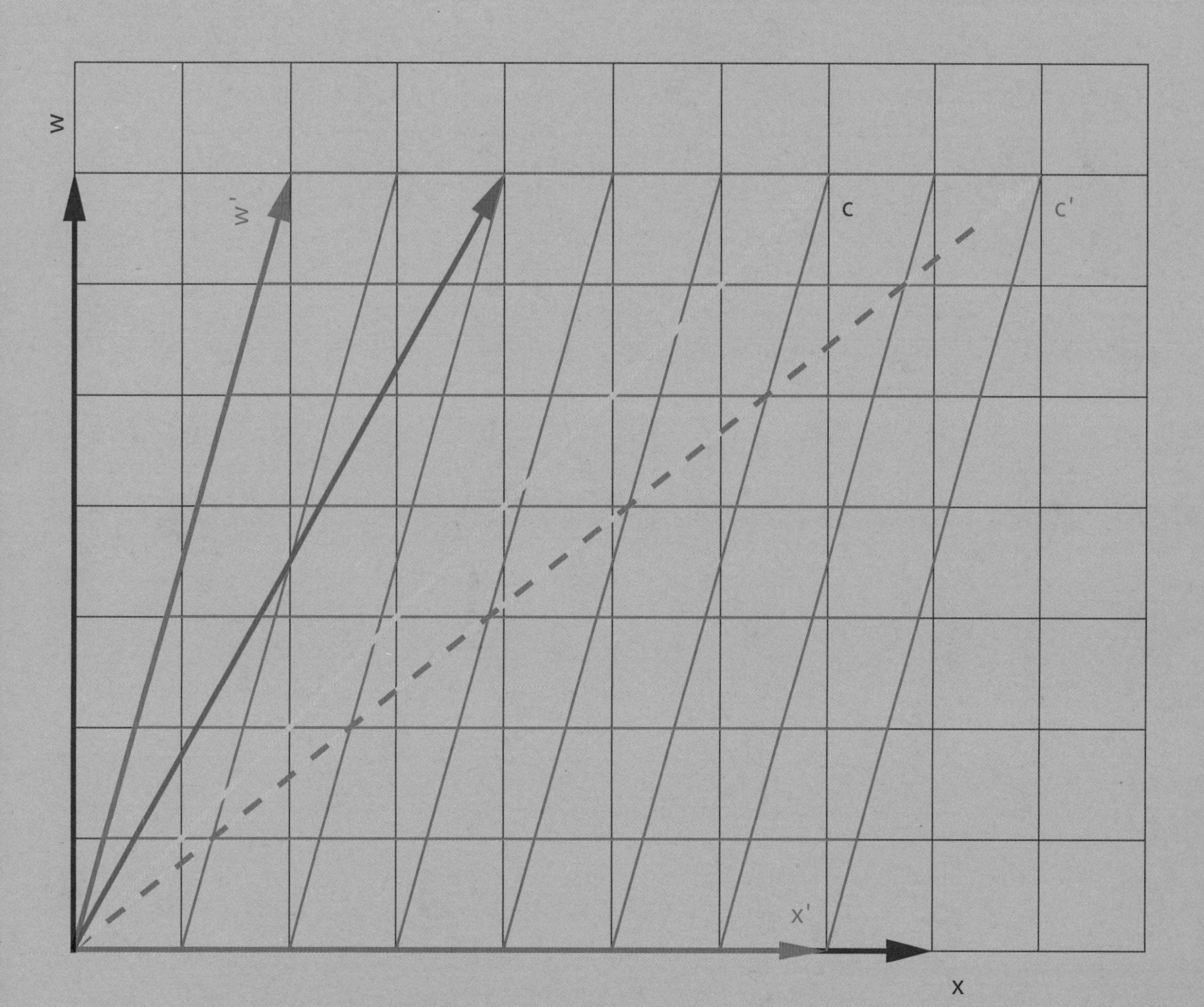
w
w'
c
c'
x'
x

Adding velocities a la Einstein

We consider a similar experiment. This time the red train moves with a velocity of $v = \frac{1}{2}c$, and the blue-eyed girl moves forward in the train also with $u' = \frac{1}{2}c$. The red world line of the train is straightforward: because it travels at half the speed of light, the two lower black double-pointed arrows must be of equal length. We also know that the velocity of light is the same in the red train and for us (the observers in black), so there is only one yellow arrow representing the light pulse. Now where should we draw the girl's blue world line?

If in the red frame something moves at half the velocity of light, then at any given time that object will have traveled half the distance the light pulse has covered in the meantime in that frame. In the train distances are measured along the red x' direction, not along the black horizontal lines. This is why the blue arrow has been drawn in such a way that the two red double pointed arrows have equal length, implying that along the x' direction the blue object at any given time has indeed traveled half the distance of the light pulse. The question we have to answer now is: to what velocity does the blue arrow correspond in the black frame, i.e. for the observers at rest? From the picture we can immediately draw some qualitative yet firm conclusions. The velocity of the blue arrow is *not* equal to the naively expected $\frac{1}{2}c + \frac{1}{2}c = c$: it is clearly *less* than the velocity of light. In fact it is quite obvious from our construction that if the blue-eyed girl runs with any velocity smaller than c in the train, she will always move with a velocity smaller than c for the observers in black as well! Conversely, if she would move with the velocity of light, then she would move at that speed for all observers, in complete accordance with Einstein's second postulate.

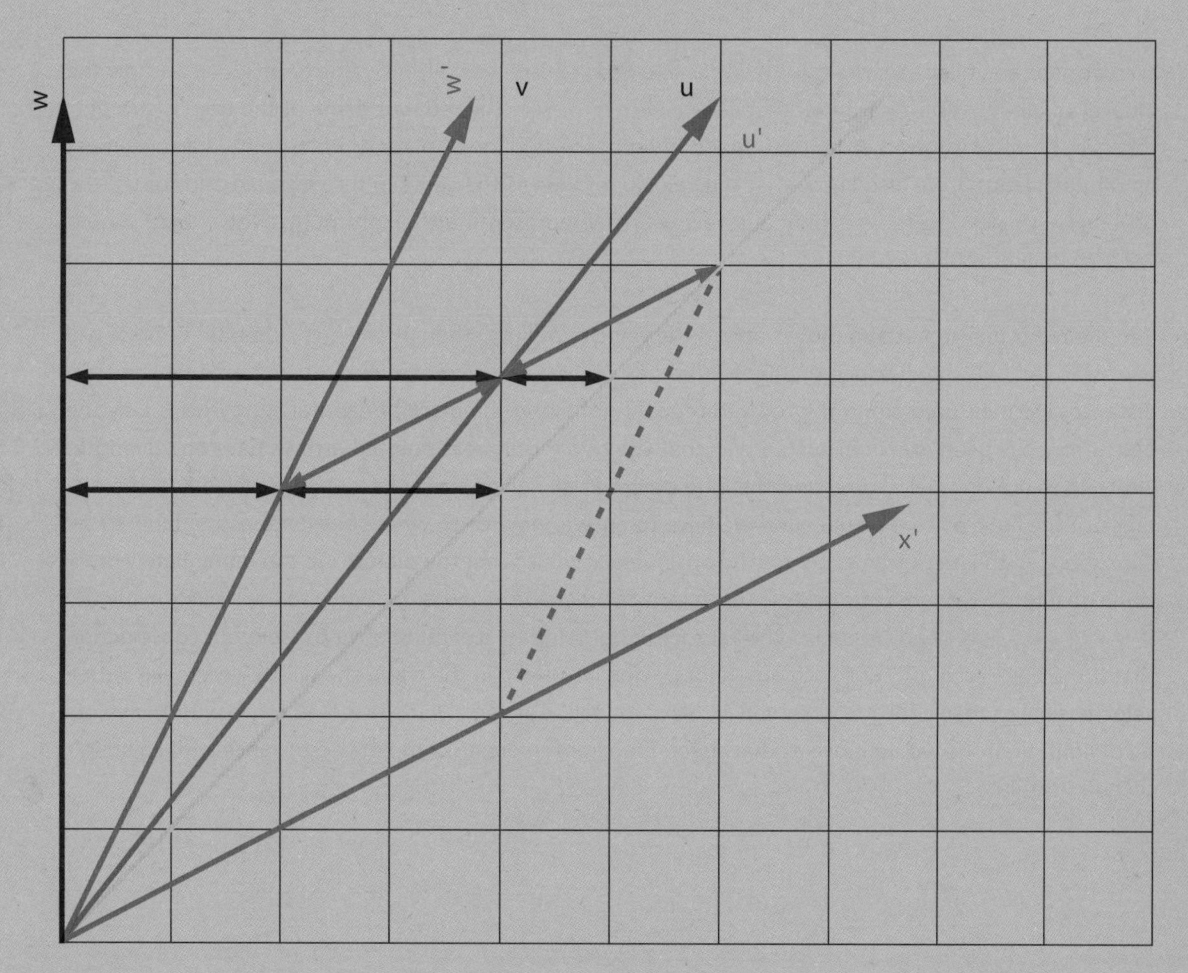

w
w'
v
u
u'
x'
x

We can go one step further and ask what would happen if the blue-eyed girl would throw a green baseball forward with a velocity smaller than c. Exactly the same reasoning can be applied, leading to the conclusion that the velocity of the ball will always be smaller than c for the black observers too. These observations lead to the startling conclusions that by adding an arbitrary number of velocities, each of which is smaller than c, one can never obtain a velocity larger or even equal to c. Briefly stated: relativity decrees that there is a maximum speed at which objects can move, and that is the velocity of light. This maximum speed is universal in the sense that it is equal for all observers. It is relatively simple to demonstrate, as we have; yet at the same time it is one of the most surprising and counterintuitive consequences of the Einstein postulates. After all, imagine a particle moving almost at the velocity of light – can we not give it a small kick to boost it past the velocity of light? The answer is negative. In chapter 6 we shall return to this apparent contradiction.

Let us finally return to the blue-eyed girl and read off from the picture which velocity she has in the black frame. We can determine the answer by comparing the lengths of the horizontal black arrows ending on the blue world line. These indicate that the velocity must be equal to 4/5 c. So the relativistic law for the addition of velocities should be consistent with ½ c <+> ½ c = 4/5 c, from which we can already draw the firm conclusion that this physical addition here indicated by "<+>" does not correspond to the standard mathematical "+" operation.

Thus far we have obtained all the important qualitative features of velocity which set special relativity so dramatically apart from the Newtonian theory. Next, you might like to learn the general formula that quantitatively describes the effects we have just discussed.

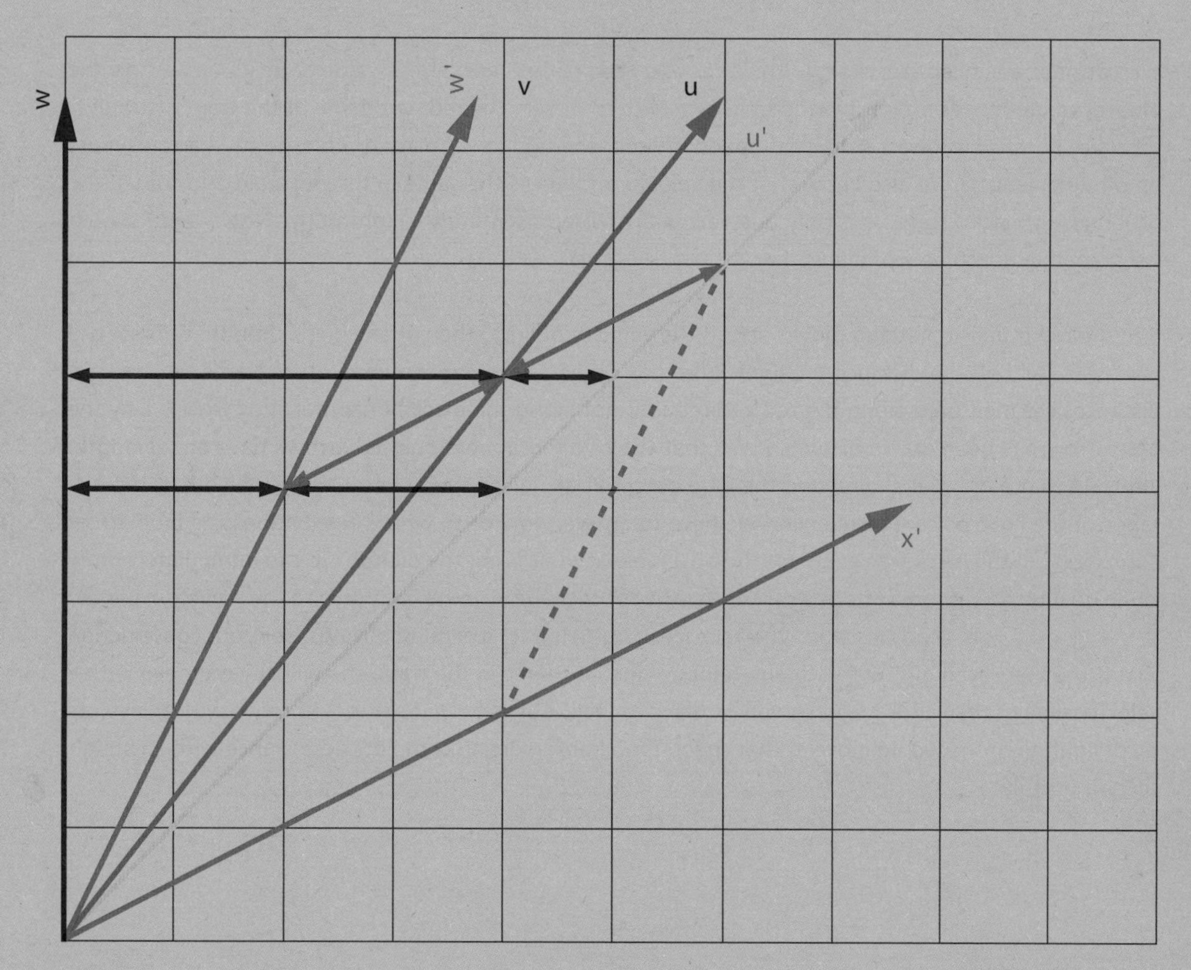
w
w'
v
u
u'
x'
x

We can go one step further and ask what would happen if the blue-eyed girl would throw a green baseball forward with a velocity smaller than c. Exactly the same reasoning can be applied, leading to the conclusion that the velocity of the ball will always be smaller than c for the black observers too. These observations lead to the startling conclusions that by adding an arbitrary number of velocities, each of which is smaller than c, one can never obtain a velocity larger or even equal to c. Briefly stated: relativity decrees that there is a maximum speed at which objects can move, and that is the velocity of light. This maximum speed is universal in the sense that it is equal for all observers. It is relatively simple to demonstrate, as we have; yet at the same time it is one of the most surprising and counterintuitive consequences of the Einstein postulates. After all, imagine a particle moving almost at the velocity of light – can we not give it a small kick to boost it past the velocity of light? The answer is negative. In chapter 6 we shall return to this apparent contradiction.

Let us finally return to the blue-eyed girl and read off from the picture which velocity she has in the black frame. We can determine the answer by comparing the lengths of the horizontal black arrows ending on the blue world line. These indicate that the velocity must be equal to $4/5\,c$. So the relativistic law for the addition of velocities should be consistent with $\frac{1}{2}c <+> \frac{1}{2}c = 4/5\,c$, from which we can already draw the firm conclusion that this physical addition here indicated by "<+>" does not correspond to the standard mathematical "+" operation.

Thus far we have obtained all the important qualitative features of velocity which set special relativity so dramatically apart from the Newtonian theory. Next, you might like to learn the general formula that quantitatively describes the effects we have just discussed.

A brief chronology of Einstein up to the "Miracle Year" 1905

1879	Born in Ulm, Germany.
1888	Enters the Luitpold Gymnasium (high school) in Munich.
1895	Leaves the Gymnasium without diploma.
1896	Obtains a diploma from Cantonal Schule in Aarau, Switzerland.
	Enters ETH (Federal Institute of Technology) in Zürich.
1900	Graduates from ETH, but cannot find a teaching job.
1902	Starts working at the patent office in Bern as technical expert.
1903	Marries Mileva Marić.
1905	March 17: Paper on existence of light quanta (photoelectric effect).
	May 11: Paper on Brownian motion.
	June 30: Paper on special relativity.
	September 27: Second paper on special relativity, containing $E = mc^2$.
	December 19: Second paper on Brownian motion.

A magic addition formula

In this section we are going to answer in precise quantitative terms the following question: If a red train travels along the platform with a given velocity v, and in the train a blue-eyed girl is running with a given velocity u', what then is the velocity u of the girl with respect to the platform? In order to find the answer, we will make use of some rather elementary plane geometry involving the properties of similar triangles.

To obtain the general expression for u in terms of u' and v, we take a series of 5 steps, exploiting the *similarity* of the two green triangles in the picture, where similarity means that the triangles have the same shape but not the same size. A typical property of two similar triangles is that the ratios of the lengths of corresponding edges are equal. I hope you are ready for some algebraic exercise. If you aren't, don't worry: you are welcome to skip the derivation (in light print) and move on to the resulting formula given below and the comments that follow it.

1. The two green triangles are similar because they can be obtained from each other by a sequence of two simple transformations. One is a reflection in a line which runs perpendicular to the yellow line and through the point where the three triangles meet, and the other is just a rescaling.

2. In the large green triangle the ratio of the two perpendicular sides s/a is equal to the distance traveled by the train, $s = vt$, divided by the distance traveled by the light pulse in the same time, $a = ct$. This ratio therefore equals $v/c = \beta$, which is independent of the particular instant in time.

3. Now the ratios of the corresponding sides of the two green triangles are equal, so $r/a = b/s$. This ratio can be determined by comparing the two long sides, which also form part of the red triangle. Arguing from the red frame, one sees that the ratio of the short to the long red side is by definition u'/c. This

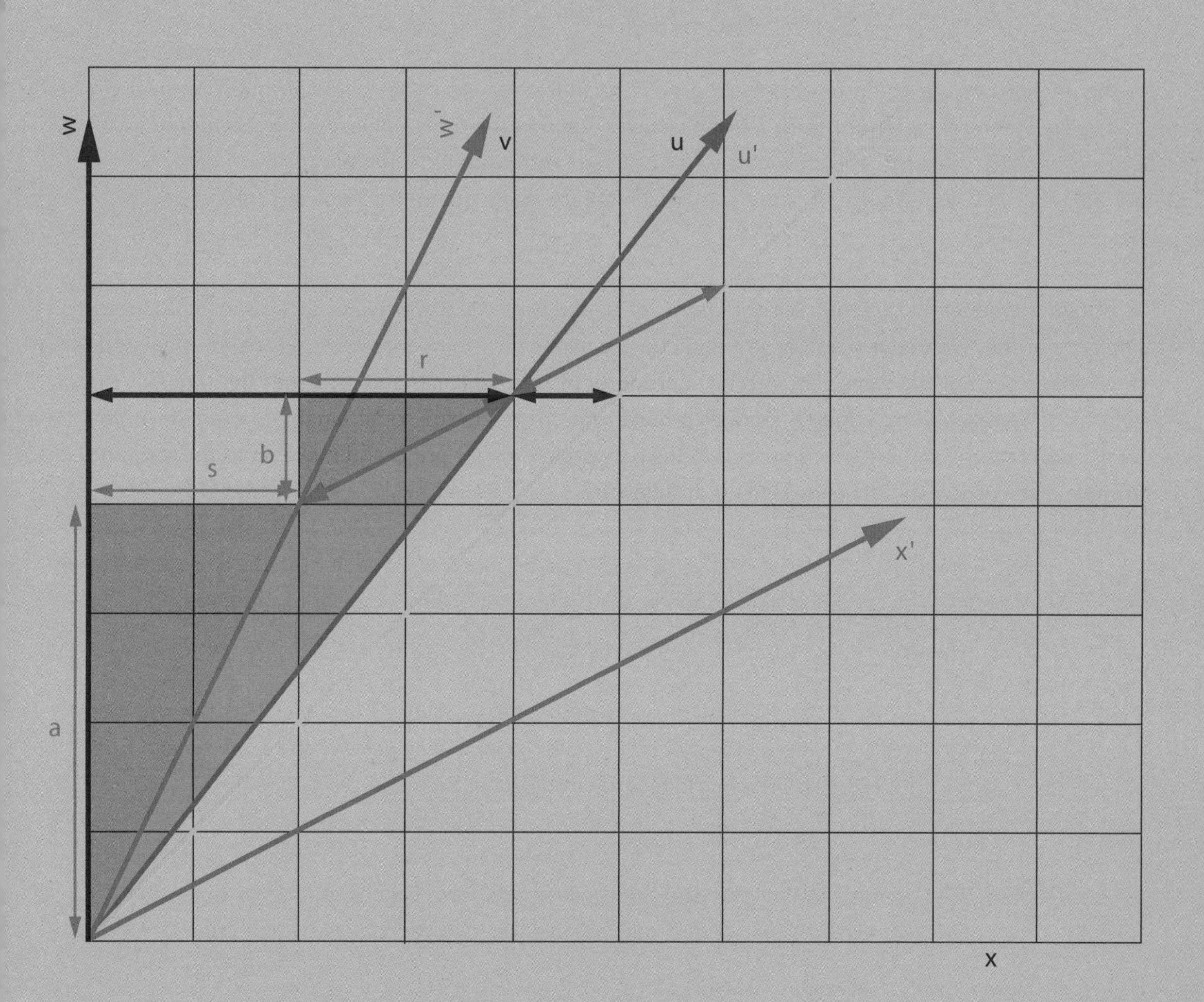
w
w'
v
u
u'
r
b
s
x'
a
x

follows from the same argument as given under 2, but now for the red frame. The long side in the red triangle represents the distance the light has covered in the red frame; it is equal to the sum of the two short red double-pointed arrows. The short side or left red double arrow, is the distance traveled in the same time by the girl inside the train. So $b/s = u'/c$ and also $r/a = u'/c$. Multiplying both sides of the first equation by s and of the second by a yields $b = u's/c$ and $r = u'a/c$.

4. The velocity u which we want to determine also satisfies a simple relation in the black frame. From an argument completely parallel to that given in step 2, we conclude that in the triangle involving the black w-axis, the black double-pointed arrow and the blue arrow, $u/c = (s + r)/(a + b)$.

5. We are done! Just substitute the expressions for b and r found in step 3 into the equation from step 4, and then use the result from step 2 that $s/a = v/c$ to reproduce the famous result first obtained by Einstein.

This is Einstein's beautiful formula for the addition of velocities.

$$u = \frac{u' + v}{1 + \frac{u'v}{c^2}}$$

Having done all this geometric labor, let us not forget to contemplate the result by checking whether it satisfies the qualitative statements made in the previous section.

- If we substitute the special values $v = \frac{1}{2} c$ and $u' = \frac{1}{2} c$, which were used in the previous example, we see that the graph did not betray us: we obtain $u = 4/5\, c$ as we did before by direct visual inspection.
- If u' and v are both much smaller than the velocity of light, so that u'/c and v/c are both much smaller than unity, we should of course recover the good old Newtonian result. With small values for u' and v the term $u'v/c^2$ in the numerator will be a lot smaller than unity, and therefore can safely be neglected with respect to the 1 that appears next to it. This leaves us with what Newton told us to expect: $u = u' + v$. This underscores the fact that Newton's physics is in a sense a special case of Einstein's and not the other way around.
- If we set u' equal to c, then the formula yields $u = c$ for *any* value of v. This is just a rephrasing of the statement that the velocity of light is the same for all observers. Even adding two times c still yields $u = c$.

Why is the addition rule for velocities just a simple addition in the Newtonian situation and so complicated with Einstein? The reason is basically that a velocity is by definition a space difference Δx (a distance) divided by (per) Δt (time elapsed). For Newton, time is universal, so Δt does not change and only Δx is affected by switching frames. In Einstein's theory both x and t transform nontrivially, which is what causes the nonlinearities in the addition formula.

Causality regained

Armed with the dictum that no velocity can exceed the speed of light, we can now return to our controversial murder case (on page 38), which we left unsolved. The fact that nothing can move faster than light implies that the effects of a certain event can never propagate through spacetime at a velocity higher than c. In the figure we have indicated what this means in our simple world of one spatial direction and one time direction. Event 1 can only causally affect subsequent events situated within the yellow wedge, which is delimited by the world lines of two light pulses moving in the positive and negative x direction (which are equal for all observers). After all, the speed at which the event's effects propagate will always be lower than c. Since in reality we are dealing with three spatial dimensions instead of one, you should actually think of the wedge as (a higher dimensional analogue of) a cone. The wedge is therefore usually called the forward or future *light cone*. If on the other hand we ask the question which events could have an effect on a given event – say event 2 – then by the same reasoning these have to be situated in its backward or past light cone, indicated in dark yellow. Note that the future and past light cones are identical in all inertial frames. The cones are universal: they are attached to an event, not to a specific observer. However, any point P located *outside* the light cones of (say) 1 may, dependent on the particular velocity of an observer passing through 1, lie in the future, the past or the present for that observer. But that ambiguity in the time ordering is innocuous, because there is no signal which can travel between point 1 and point P. There can be no causal relation between events at points 1 and P.

Now if we return to our causality problem on pages 38-39, we see that the events 1 and 2 are outside each other's light cone. Causality is thus rescued from its demise. In case it bothered you: naughty Nigel cannot have killed Auntie Augusta!

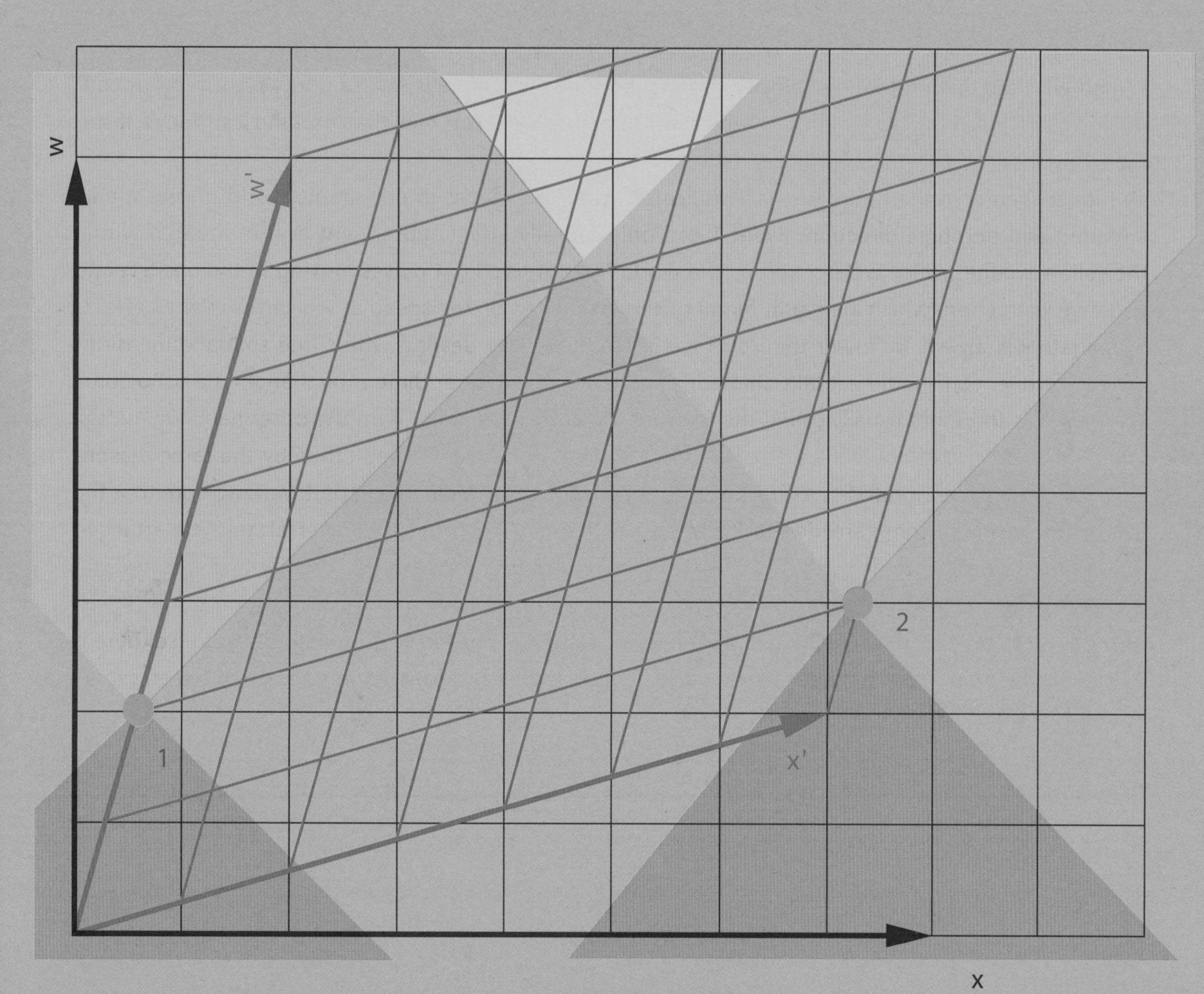
w
w'
1
2
x'
x

4 Dilations and contractions

Everything should be made as simple as possible, but not one bit simpler.

Excuse me, can you tell me what time it is?

Simultaneity is relative: which events occur at equal times depends on your frame of reference, which is determined by your velocity. Looking at the figure one might ask the following question: What time is it at the point w'? For the black observers w' is simultaneous with $w = 5$ units, and for the red observers it is simultaneous with $w = 3.3$ units. Another conundrum, it seems. We should not be surprised – after all simultaneity is relative, and in the above example all observers are referring to the black watch. The interesting question of course is what time it is on the red watch at w', and how this is related to the time assigned to the same event in the black frame. We can be sure about one thing: supposing the red observer has set her watch to zero at the origin, it will indicate one particular time at w'. To find out what time, we shall make deliberate use of the relativity postulate.

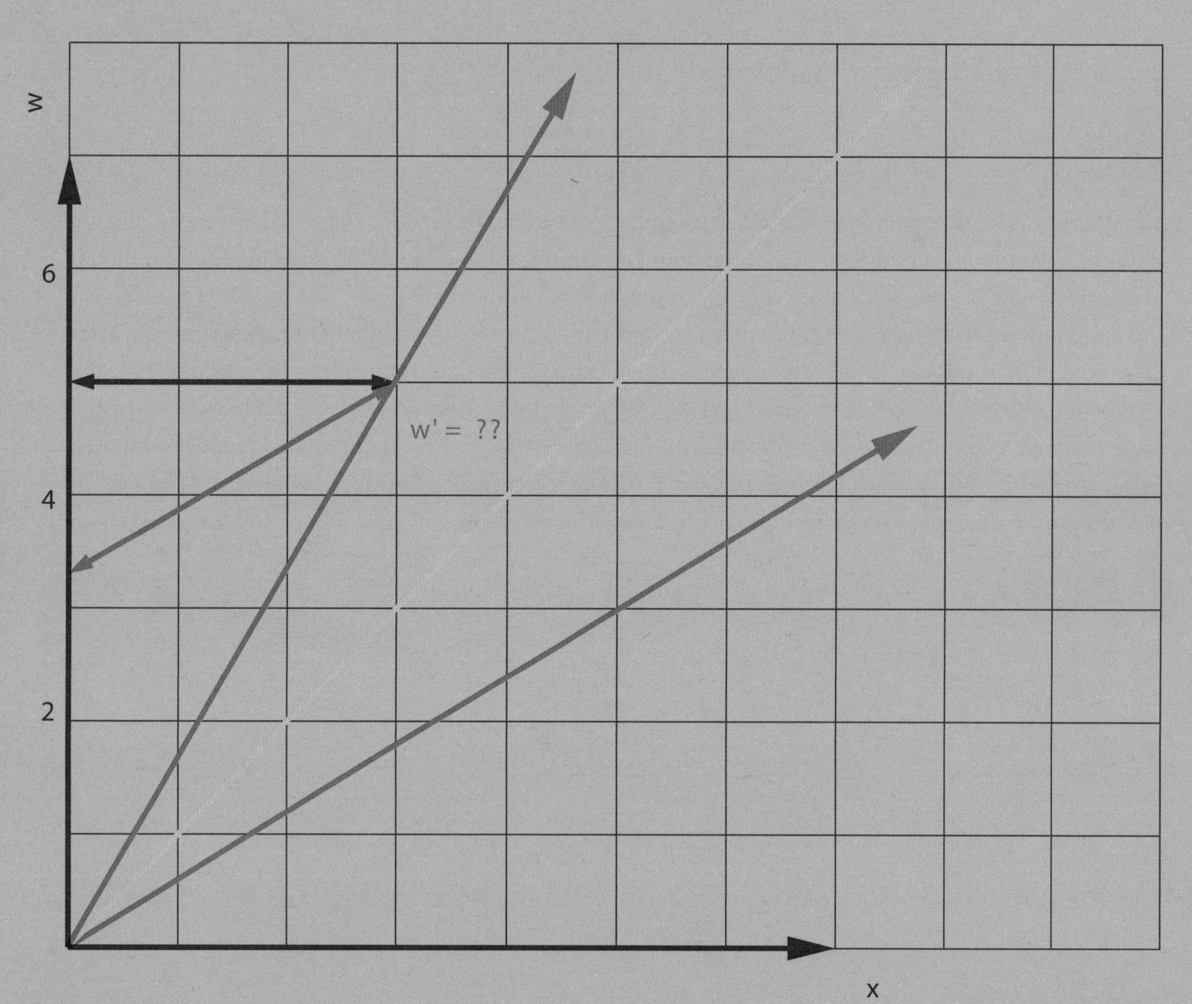
w
6
4
2
w' = ??
x

Time dilation

The problem raised in the previous section can be resolved by applying the relativity principle to the clock rates of the two inertial observers moving with a velocity v with respect to each other. The two observers in question carry identical clocks set to zero at the origin and each mark the time units on their own world line. We shall see that the clock rates must differ by some factor γ (gamma), which depends on the relative velocity v, or rather on the dimensionless parameter $\beta = v/c$. We also know that the rates should become equal as v tends to zero.

Referring to the figure, let us set $w' = \gamma w^*$. Now relativity dictates that it should also be true that $w = \gamma w'$, since there is only one relative velocity and the situation should be entirely symmetrical for the two observers. Substituting the expression for w' into that for w, we obtain the relation $w = \gamma^2 w^*$, where both w and w^* refer to the same clock in the black frame. There is one firm conclusion we can already draw at this point: because it is obvious in the figure that w is larger than w^*, it follows that γ^2 and therefore also γ itself has to be larger than unity. From $w' = w/\gamma$ it follows that w' must be smaller than w. This means that *a moving clock runs more slowly*; a result that is rather bizarre and very important. Indeed, from the figure one would rather infer that $w' > w$, and the fact that that is not true literally means that we have to rescale the units along the oblique axes of the moving frame.

If you are a diehard, of course you now want to know how much more slowly the moving clock is running. Again this can be calculated using some elementary plane geometry, as we will now demonstrate.

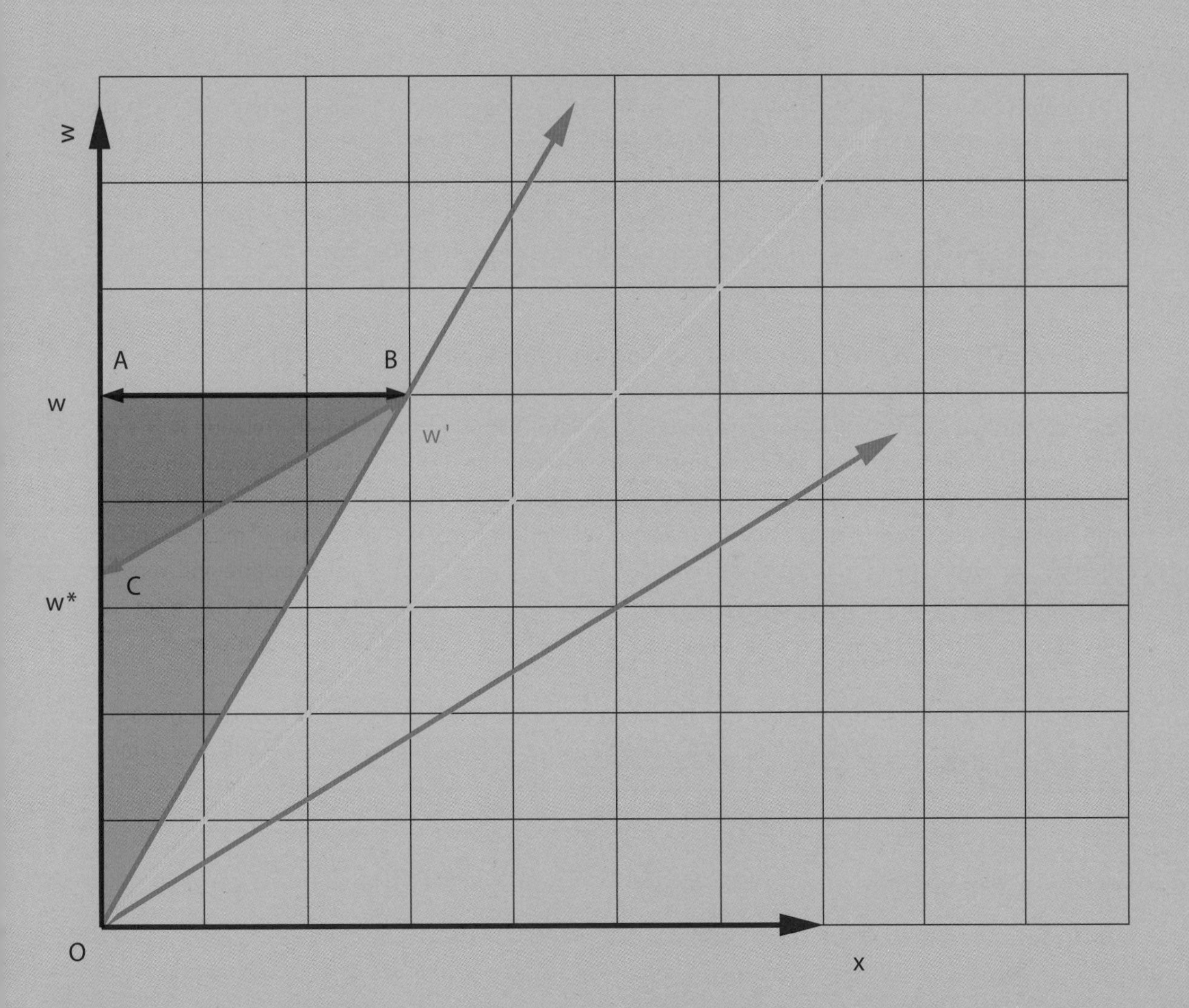
w
A
B
w
w'
C
w*
O
x

If you do not want to plough through the detailed steps of the derivation, you can go straight to the resulting formula and the comments that follow it.

In the figure on the previous page there are two green rectangular triangles: a big one, ABO, and partly overlapping it a smaller one of a darker color, ABC.

1. These triangles are again similar, so the ratios of corresponding sides are equal. From this observation it follows that AB/AO = AC/AB.

2. First note that $AB/AO = v/c = \beta$ and $AO = w$, so that $AB = wv/c = \beta w$. In the picture we can also directly see that $AC = w - w^*$. Putting these expressions back into the equation from step 1, we get $\beta = (w - w^*)/\beta w$.

3. We can solve for w by multiplying both sides of the equation by βw, moving all terms containing w to one side and taking w outside the brackets. We then obtain the formula $w = w^*/(1 - \beta^2)$.

4. Recalling that $w = \gamma^2 w^*$, we find that the scaling factor γ is given by the (positive) square root of the fraction $1/(1 - \beta^2)$.

We conclude that the relation between the clock rates is given by:

$$w' = w\sqrt{1 - v^2/c^2}$$

Let us now comment on some of the salient features of this remarkable formula. We see, as argued before, that indeed w' is always smaller than w because the factor under the root is always smaller than unity (because v is of course smaller than c). It is comforting to see that $w' = w$ if $v = 0$, and maybe less comforting to see that w' approaches 0 as v gets close to c. In other words, a clock that moves with the velocity of light does not run at all! In that unique frame the notion of time is lost: The "oblique" frames for moving observers we have been drawing so nicely collapse to a single line, on which the distinction between space and time is entirely lost.

Note: All along, we have been using good old Euclidean plane geometry for our calculations. One might worry about whether it is correct to use the notions of Euclidean similarity in the present context, where we are comparing different spacetime frames. Are the rules of Euclidean geometry valid in the spacetime plane? As a matter of fact, not really. The reason is not that the red frame appears as oblique, but rather that the units along the red axes have to be rescaled. However, the corresponding sides of the triangles we have been comparing always belonged to the same frame. In the ratios of edges of the same color the scaling factor will cancel out, and hence these ratios can be equated without harm.

We have seen that spacetime diagrams are an extremely powerful aid in order to gain understanding of relativity. Even so, they never make the equivalence of inertial frames manifest directly, in a *single* picture. But a similar asymmetry between frames also exists in the algebraic framework, where one uses a single formula like the one we have obtained for time dilation.

The Doppler effect

If we put a brass band playing "Oh when the Saints..." on a truck, the pitch of the tune will be higher when the truck is moving towards us, and lower when the truck is moving away. This change in tone height or frequency depending on the relative velocity of the source and observer is called the *Doppler effect*. It applies to all wave phenomena, water waves and sound as well as light. And in all cases the effect depends on the difference in speed between the source and the observer.

In the figure we have depicted a moving light source that flashes with frequency f_s. As you can see, the observer at rest receives the light signals with a different frequency, and the question is what frequency f_o she will measure. The frequency is just the number of pulses per second. So from the figure we see that $f_s = 4/w'_0$ and that $f_o = 4/w_1$. This means the ratio of the observed to the emitted frequency is $f_o/f_s = w'_0/w_1$. The time dilation formula tells us that $w_0 = \gamma w'_0$. From the figure we can read off that $w_1 - w_0 = \beta w_0$, because this distance equals the length of the horizontal arrow, which is the distance traveled by the source in time w_0 at velocity βc. We conclude from this that $w_1 = (1 + \beta) w_0 = (1 + \beta) \gamma w'_0$. Hence the relativistic Doppler effect is given by:

$$\frac{f_o}{f_s} = \frac{1}{(1+\beta)\gamma} = \sqrt{\frac{1-\beta}{1+\beta}}$$

Brainteaser: Show that we obtain the nonrelativistic case by setting $\gamma = 1$ in the formula above, which may then also be used for the brass band if in the expression for β we replace c by the speed of sound.

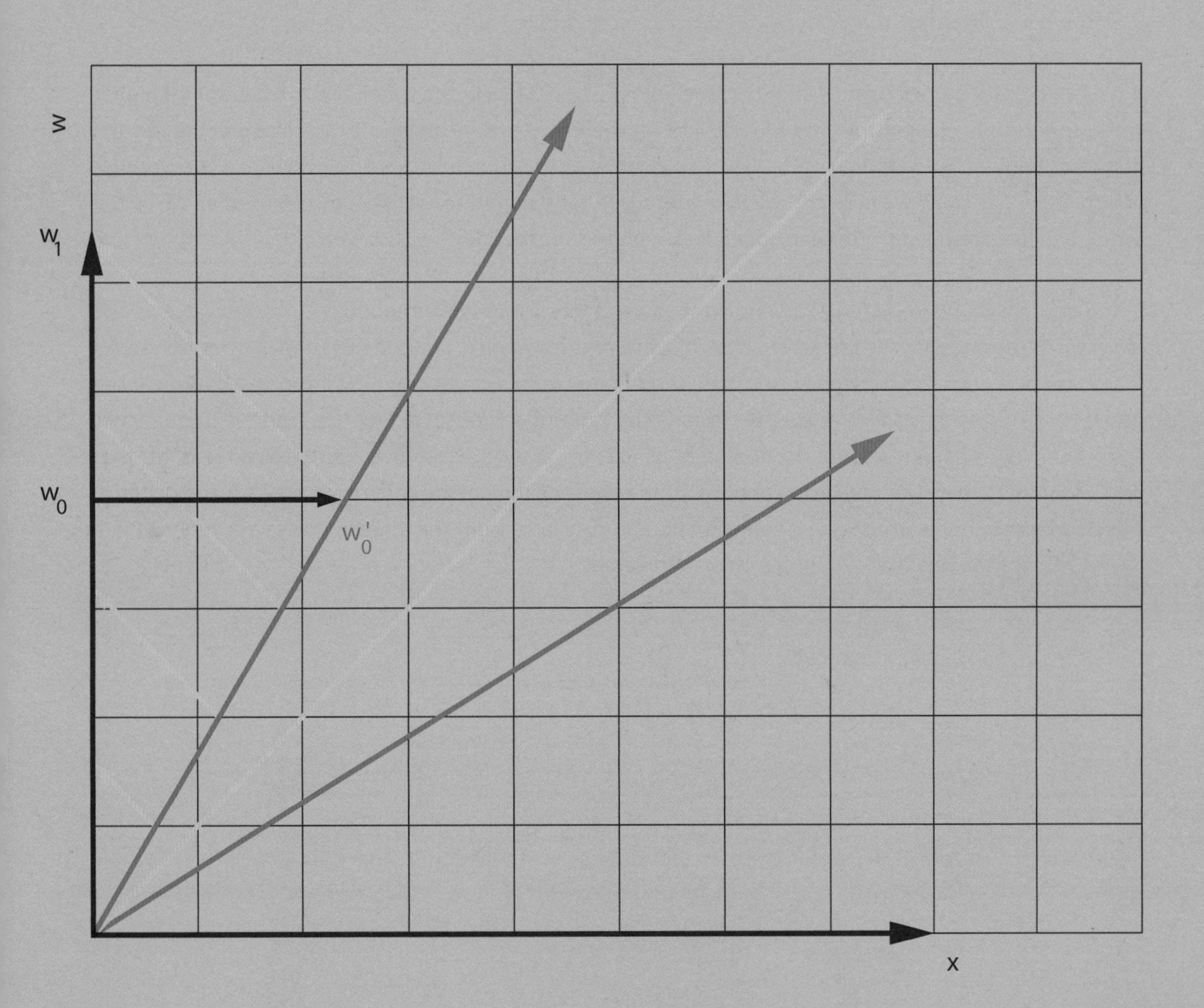
w
w_1
w_0
w'_0
x

The twin paradox

The twin paradox shows that the time dilation effect, the fact that moving clocks run at a slower rate, is real. Time dilation as a real physical effect appears to be yet another paradox. After all, doesn't relativity say that motion is relative? If A's clock runs more slowly than B's because A is moving with respect to B, shouldn't we also require that B's clock run more slowly than A's, as B is moving with respect to A just as well? This is the paradox underlying the following thought experiment.

Nora and Vera, two identical twins, are given identical, perfectly calibrated clocks. Vera then goes on a space trip, moving through the galaxy at great velocity, to return home after a long journey. Nora stays at home. At a certain time Vera comes back. Because she has been moving, her clock has been running more slowly and therefore for her less time has elapsed since she left. She will find her sister much more aged then herself. Depending on the length of her trip and the relative velocity she traveled at, she may even find that Nora has died long ago! Now that is drama.

Is this brilliant fiction or harsh reality? And if it is for real, how can we reconcile this asymmetry with the basic postulate of relativity? That is the question! Well, it is true. The asymmetry becomes clear if we look carefully at the schematically depicted travel adventure in the figure. The observer moving along the black w-axis must clearly be Nora, who stays home at rest waiting for her sister to return. Vera's rather boring journey in the red space ship consists of two symmetric parts: first she moves away with a velocity v, and then she turns around (instantly) to go home with velocity $-v$.

Since the time dilation only depends on the velocity squared, her clock runs more slowly by the same amount on the way out and on the way back. According to the time dilation formula from the previous section, when according to Nora's clock at rest say $t_1 = 30$ years have elapsed, for Vera only t_1' years have passed, where $t_1' = t_1\sqrt{(1-\beta^2)}$. By setting v appropriately, the traveling twin can make t_1' as small as she

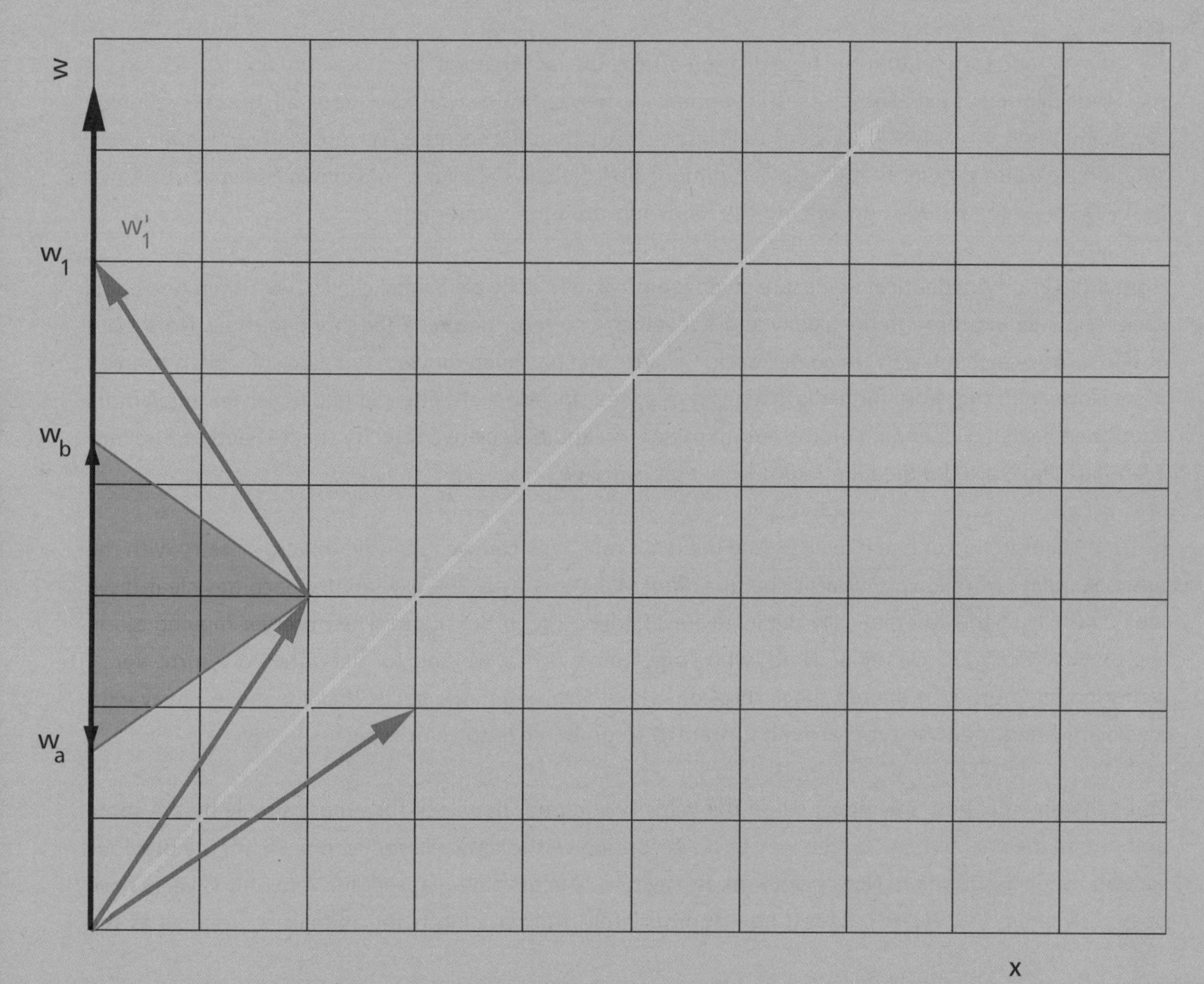
w
w'_1
w_1
w_b
w_a
x

wishes. For example choosing $v = 4/5\ c$ yields $t_1' = 3/5\ t_1 = 18$ years!

The picture reveals where the asymmetry arises. Just before the turning point, Vera considers w_a to be simultaneous to her time, but an infinitesimally small amount of time later she sees w_b as simultaneous. So somehow she has jumped instantaneously from w_a to w_b – or more realistically, if the curve is smoothened out a little, she sweeps extremely rapidly through the time period between w_a and w_b. This is not a relative statement, because Vera's velocity *changes*. She goes through a very rapid deceleration, and that is something that she can objectively determine, just as you feel it right away when in a car that suddenly brakes. Her sister Nora does not experience any of those decelerations and this is where an objective asymmetry arises – and that difference is what resolves the paradox.

Clearly this dramatic asymmetry is unavoidable if one sister is to stay at rest and we insist that the two sisters have to meet again to compare their actual ages. Some experts therefore say that the time difference is caused by the acceleration and is not just an effect due to special relativity. Yet the overall effect depends directly on the relative velocity of the two observers and the duration of the trip. We can approximate this effect arbitrarily the accumulation of smaller segments where the traveling twin moves with different constant velocities in relation to the sister at rest. Furthermore the smoothening of the sharp corner has an effect that has nothing to do with the length of the trip and can therefore be made arbitrary small.

Most important is the fact that the twin paradox is an entirely real physical effect, of which there are direct experimental verifications. In 1971 experiments have been performed where a very accurate atomic clock was sent around the earth in a jet plane with an average velocity of about 600 miles per hour. The result was a tiny but significant difference in time measurements with the identical clocks that stayed behind in the lab, in full agreement with Einstein's formula. The calculation of this special case also serves to illustrate the paradoxical general property that if two people travel along arbitrary world lines that part and meet again, then the traveler who took the "longest" world line will be the youngest.

The effect of time dilation can be verified in a very basic way using unstable elementary particles like muons, which decay spontaneously and have a finite (average) lifetime. This lifetime has indeed been found to depend on the velocity of the decaying particles in relation to the laboratory in which their lifetime was determined. These experiments provide a very precise confirmation of Einstein's prediction. They also underline the fact that the effect does indeed belong to the realm of special relativity, as in this experiment there are no accelerations and yet the lifetime differs in the different frames. This can be done because in fact time and distance measurements are made in the lab, which are compared to a time measurement in the frame of the decaying particle (for which the particle itself acts as clock). So time dilation effects as a consequence of special relativity, and of the relativity of simultaneity in particular, are just as real as the law of nature that states that a particle will accelerate when a force is exerted on it.

Brainteasers: 1. Imagine that Nora and Vera both send light signals to each other at a rate of one per second. Draw the light rays in the diagram and discuss how the sisters perceive each other's sequence of signals. Now the "real" asymmetry will show up.
2. Draw a diagram for the experiment with the decaying muon, to show that it allows for a verification of the time dilation effect. Assume the muon's velocity is ½ *c*.

Lorentz transformations

We have been talking all along about different frames of reference, which are related to different coordinate systems like the black or the red grid. With two coordinate systems belonging to two sets of inertial observers (moving at a constant velocity with respect to each other), there is an important general question one might ask.

Given an event P, which has coordinates (w,x) in the black (rest) system and coordinates (w',x') in a frame moving with velocity parameter $\beta = v/c$, what is the general relation between the coordinates (w,x) and (w',x') in the two frames? In other words, we are looking for the expression of w and x in terms of w' and x' (or the other way around). If somebody tells us the "when and where" of some event in one frame, we can then directly calculate its "when and where" in the other frame. Not surprisingly, this relation will depend on β. It can be obtained from the pictures we have been drawing, using similar geometric arguments to the ones we have been using, and that is what we are going to do now.

The relations we are looking for are the renowned *Lorentz transformations*, which allow you to transform or translate expressions from one inertial frame to another. We have already encountered a simple example of this in the section on time dilation, which involved a relation between w' and w. A derivation of the relations between the two coordinate systems is given on the next page. It can of course be skipped if you are merely interested in the result and its implications.

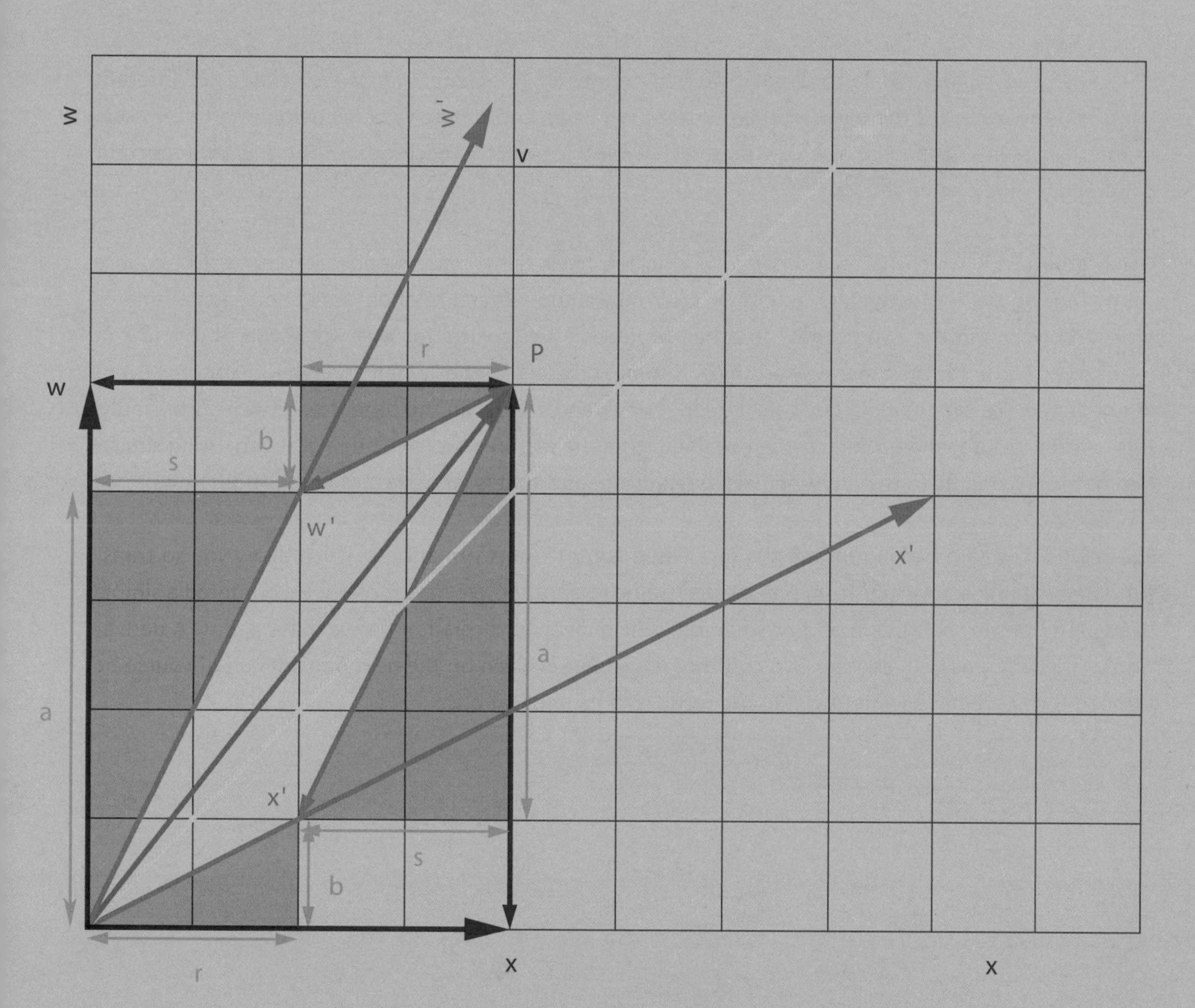
w
w'
v
r
P
w
b
s
w'
x'
a
a
x'
s
b
r
x
x

We can quite easily obtain the desired relations from the figure, using the proportionality and in certain cases equality of the edges of the green triangles.

1. The event P is marked in blue, and along the red coordinate axis we have indicated that it has coordinates w' and x'. Along the black axes we have marked its coordinates in the rest frame as w and x.

2. From the rectangular triangles in the figure we learn that $w = a + b$ and $x = r + s$.

3. We have made extensive use before of the fact that $s/a = b/r = v/c = \beta$. In the section on time dilation we also demonstrated that $a = \gamma w'$, where γ stands for the expression given on page 56. Substitution into $s/a = \beta$ yields $s = \beta\gamma w'$. From the similarity of the big and small triangles it also follows that $r = \gamma x'$ and $b = \beta r = \beta\gamma x'$.

4. Now all we have to do is substitute some of the expressions derived in step 3 into those from step 2. We directly obtain $w = a + b = \gamma w' + \beta\gamma x'$, and similarly we get $x = r + s = \gamma x' + \beta\gamma w'$. These simple transformation rules for switching from spacetime coordinates (w', x') to coordinates (w, x) do indeed exhibit all the expected symmetry properties suggested by the picture. They also show the correct behavior if we take the limit $v \rightarrow 0$ (so that $\beta \rightarrow 0$ and $\gamma \rightarrow 1$).

We have now arrived at the well-known Lorentz transformation rules:

$$w = \gamma w' + \beta\gamma x'$$
$$x = \gamma x' + \beta\gamma w'$$

This is a fundamental result of great generality.

There is a mathematical property to the transformation rules. This transformation is called linear, because the new w and x are expressed as linear combinations of the old w' and x'. (This means no higher powers are involved.) The coefficients β and $\beta\gamma$ are of course dependent on the relative velocity of the two observers. Note also that the nonrelativistic limit is the equally linear Galilean transformation: $w = w'$ and $x = x' + \beta w'$.

The linearity of the transformation reflects a property of spacetime itself which we have tacitly assumed. We have stated that the properties of the frames of reference were *homogeneous* meaning that they do not depend on where or when you are, that empty spacetime looks the same around any point. It allowed us to arbitrarily choose an origin. This can be shown more explicitly as follows. Assume we choose a point (a,b) as a new origin, this corresponds to the point (a',b') in the moving frame. Homogeneity is the requirement that the transformation from (w',x') to (w,x) is the same as from $(w'-a',x'-b')$ to $(w-a,x-b)$, and it leads to the condition that the transformation is linear.

This linearity is a very pleasant property, it ensures for example that if we apply two or more of such transformations sequentially, the combined effect is again a linear transformation. For example first we transform from the frame of people on the platform to the frame of the red train with parameter $\beta = \beta_1$ then we transform from the train to the frame of the blue-eyed girl with a parameter β_2. If we perform the transformations after each other you can show that the overall result is the same as a single transformation with a parameter β_3 given by the Einstein formula for the addition of velocities of page 48, so, $\beta_3 = (\beta_1 + \beta_2)/(1 + \beta_1 \beta_2)$. Hidden under the complicated nonlinear addition formula are the simple linear Lorentz transformations.

In many textbooks the Lorentz transformations are actually taken as the starting point for an explanation of relativity. This makes sense from a historical point of view, because of the remarkable fact that these transformation formulas had already been written down by the Dutch physicist Hendrik Antoon Lorentz around 1900 prior to the emergence of relativity. They followed from his analysis of Maxwell's theory, the set of equations that give a unified description of electromagnetic phenomena. Lorentz made the remarkable discovery that the Maxwell equations look exactly the same if one changes the coordinates from primed to unprimed variables according to the above transformation. In physics and math jargon the appropriate statement is that the equations are *invariant* under the Lorentz transformations.

It is fascinating to realize that the fundamental equations of relativity had somehow already been written down before Einstein got to them. Apparently the problem here was not so much finding the right answer as asking the right question about their meaning. Indeed, the initial interpretation of the invariance was entirely different. It was believed that the Maxwell equations took their simple and beautiful form only in a particular rest frame, which was at rest with respect to the "ether ", an elusive substance that was supposed to fill all of space. It was the medium that was believed to be necessary for the transmission of electromagnetic waves (such as light or radio waves). The profound and radical turn in interpretation due to Einstein was that there was no such thing as an ether, and as a consequence no such thing as a "preferred" frame of reference. This view was in line with the findings of a famous experiment, actually performed prior to the advent of relativity, by Michelson and Morley, in which they demonstrated that light propagated at the same speed in all directions. This contradicted the widely accepted idea that the earth was in motion with respect to the ether. It provided strong experimental support for Einstein's second postulate, though it is not entirely clear to what extent Einstein was fully aware of this.

The crucial observation was that while Newtonian mechanics was "invariant" under the Galilean transformations, which we just mentioned, Maxwell's theory of electromagnetism was invariant under the very different set of Lorentz transformations. For the relativity postulate to hold, namely that all physical equations should look the same for any observer moving at a constant speed in relation to another observer, one of the two theories would have to be changed. It was this insight which brought Einstein to his bold revision of Newtonian mechanics, once thought to be invincible, while he left Maxwell's theory untouched.

Brainteasers: 1. Show that the relations giving x' and w' in terms of x and w can be obtained from the above relations by replacing β by $-\beta$. Exactly what one would expect from relativity.
2. Show that two subsequent Lorentz transformations with parameters β_1 and β_2, amount to the same as a single transformator with parameter β_3 given by the Einstein formule.

Does the pole fit in the barn?

Looking at the expressions for the Lorentz transformations, you may have noticed that space and time appear on an equal footing in relativity. As we have already encountered the physical effect of time dilation, it is natural to ask whether there exists a similar physical effect associated with the space coordinate. Indeed, there is such an effect and it is called the "FitzGerald-Lorentz contraction". Briefly stated it asserts that the length of an object moving at constant speed will be observed as contracted (in the direction of motion).*

Let us briefly illustrate this effect in the context of another paradox which arises when one wants to answer the question whether a pole fits in a barn. The paradox involves a barn at rest and a pole moving through it. For the observer at rest the pole is contracted, and he observes that it just fits in the barn. For an observer who is moving quickly while carrying the stick, the barn is contracted and the stick is not, so according to her the stick does not fit in the barn. How do we decide who is right? Does the pole fit or does it not, that's the question.

In the figure we have depicted the situation. Firstly, there is the black rest frame. The light green area is the barn, being at rest; the two black arrows pointing up correspond to the world lines of the front and back doors of this (one-dimensional) barn. The pole, the double-pointed arrow, is moving with a constant velocity in the positive x-direction and it is at rest with respect to the red frame. The two red arrows pointing diagonally upward on the right represent the world lines of the endpoints of the pole.

* In case you are wondering, the corresponding formula is: $x = x'\sqrt{(1 - \beta^2)}$. Note that the x and x' appear in the opposite places compared to the time dilation formula on page 56.

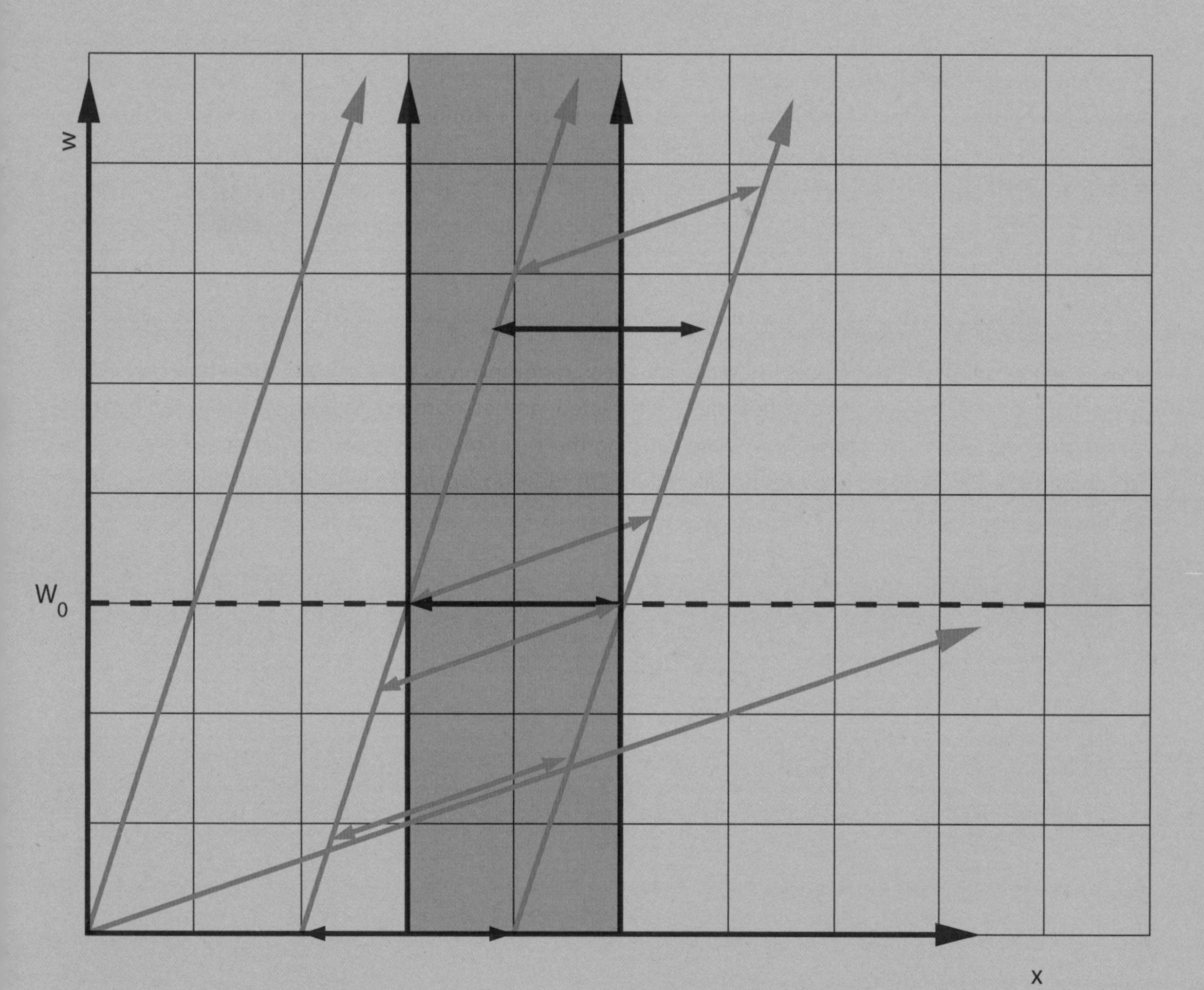

w
W_0
x

In the figure the resolution of the paradox is clear. In the black frame the length is measured along horizontal equal time lines, and we see that the pole fits exactly in the barn: at a given instant $w = w_0$, both endpoints are inside the barn. For the red observers the story is very different: at time w_0, when the front end of the pole reaches the back door, the other endpoint of the pole has not yet entered the barn. The moving observer concludes correctly that the pole does not fit in the barn. The clue is thus that length measurements by definition involve the notion of simultaneity. As this is frame-dependent, consequently, so is any statement comparing the lengths of objects moving with different velocities.

The answer to the question "Does the pole fit in the barn or not?" must be: "That depends." Not only on the pole, but also on the observer. Both observers spoke the truth, or at least, they spoke *their own* truth.

Brainteaser: Consider the following thought experiment, suggested by Taylor and Wheeler: A train is moving along a wall on which a blue line has been painted, exactly two meters above the ground. In the train a man with a paintbrush is leaning out the window. He intends to paint a red line on the wall, also exactly two meters above the ground. Will the red line end up below or above the blue one? Argue that if there is more than one space dimension, the dimensions perpendicular to the direction of motion are *not* contracted.Start out by assuming that the vertical dimension of the moving frame will also be contracted, and use the relativity postulate to argue that this leads to a contradiction.

Einstein the man

Einstein was the freest man I have known. By that I mean that, more than anyone else I have encountered, he was the master of his own destiny. If he had a God, it was the God of Spinoza. Einstein was not a revolutionary, as the overthrow of authority was never his prime motivation. He was not a rebel, since any authority but the one of reason seemed too ridiculous to him to waste effort fighting against. He had the freedom to ask scientific questions, the genius to so often ask the right ones. He had no choice but to accept the answer. His deep sense of destiny led him farther than anyone before him. It was his faith in himself which made him persevere. Fame may on occasion have flattered him. He was fearless of time and to an uncommon degree fearless of death. I cannot find tragedy in his later attitude to the quantum theory or in his lack of finding a unified field theory, especially since some of the questions he asked remain a challenge to this day – and since I never read tragedy in his face. An occasional touch of sadness in him never engulfed his sense of humor.

Abraham Pais
in his Einstein biography
Subtle is the lord...

5 A geometric interlude

Do not worry about your difficulties in mathematics. I can assure you mine are greater still.

The spacetime interval

We have seen that there are two aspects in which the coordinate grids we have been drawing differ. The first difference is that in one of them the axes are perpendicular (the black frame) and for the others, the time and space axis are oblique and not perpendicular. The second difference is that we have to rescale the units along the axes by the velocity-dependent factor $\gamma = 1/\sqrt{(1-\beta^2)}$.

Let us now consider the collection of all observers who move through the origin at time zero, but at different velocities. We ask all of them to mark on their world line the event where for them, on their clock, a given fixed time – say, s units – has elapsed. One might wonder what the resulting set of events would look like in a spacetime diagram. The time dilation formula tells us that $(w')^2 = (1-\beta^2)w^2$, so setting $w' = s$ we obtain the expression $(1-\beta^2)w^2 = w^2 - (\beta w)^2 = s^2$. We also know that in the rest frame the distance x the moving observer has moved away from the origin, equals $x = vt = vw/c = \beta w$. So the result is that the points where each observer measures his time equal to s, form a curve in the (x,w)-plane described by the following, strikingly simple formula:

$$w^2 - x^2 = s^2$$

What does this curve look like? Well, you might be familiar with the same equation, only with a plus instead of a minus sign. Then the curve would be a circle with radius s, centered at the origin. With the minus sign we do not get a circle, but another celebrated mathematical curve called a *hyperbola*.

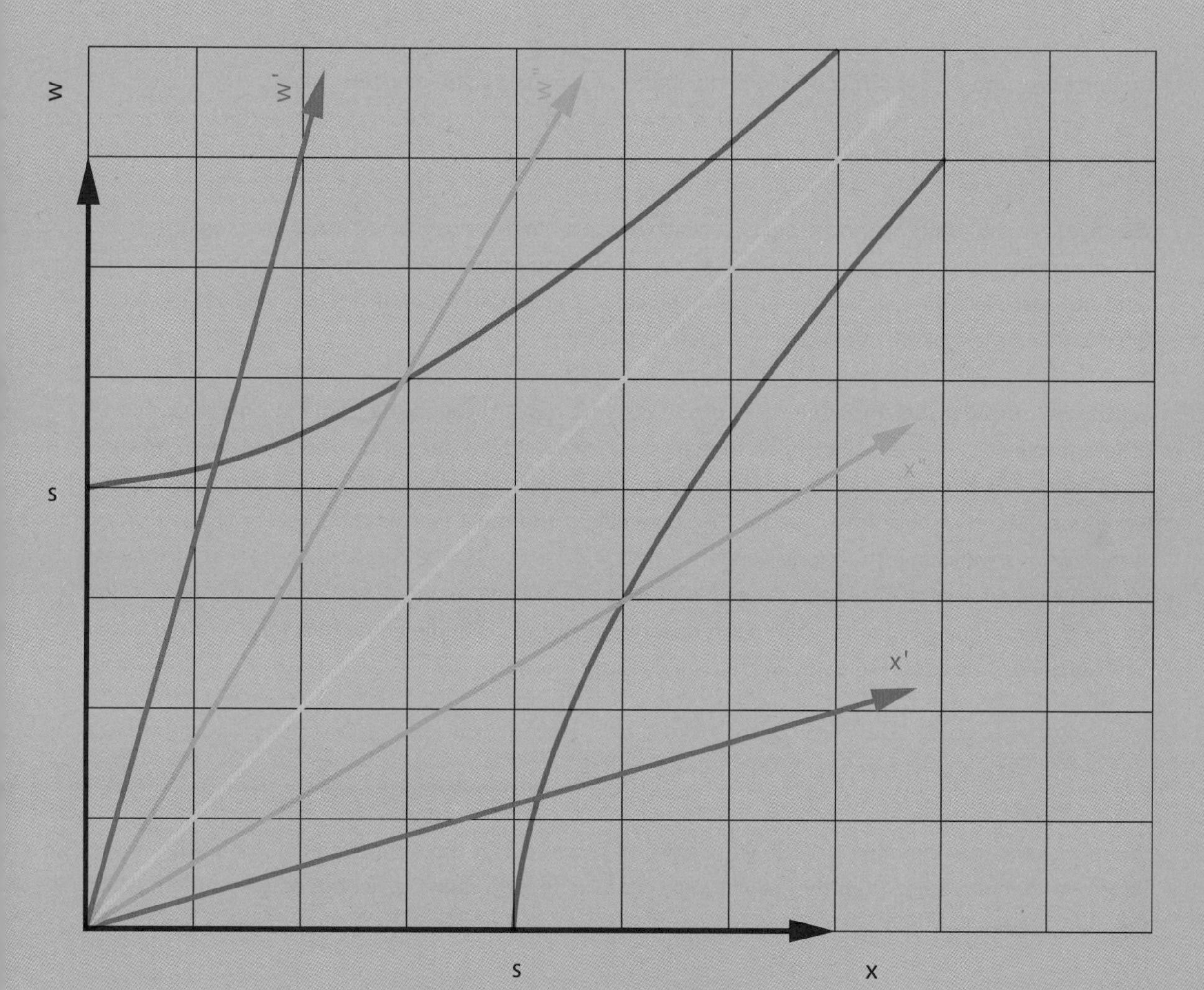
W
W'
W"
S
X"
X'
S
X

Just like a circle is completely characterized by its radius, our hyperbola is characterized by its intersection point with the w-axis, which is the number s. Using the formula one can just plug in values for x, calculate the corresponding w values, and then plot the points in the spacetime diagram. Connecting those points one gets curves like the dark blue one in the figure on the previous page. The "horizontal" hyperbola is characterized by $s = 4$; it intersects the black, red and light blue w-axes in points where $w = 4$, $w' = 4$ and $w'' = 4$ successively.

Of course we can play the same game with meter sticks, where the different observers mark a distance $x' = s$ at time zero on their world line. This yields the formula with x and w interchanged, which amounts to replacing s^2 by $-s^2$ in the formula above. Drawing the corresponding curve yields the other blue hyperbola in the figure, which intersects the x-axis in point s. The hyperbolas are often called the *spacelike* and the *timelike* hyperbola. The case in between, with $s = 0$, is somewhat degenerate: the hyperbolas turn into the lines $w = +x$ and $w = -x$, the world lines of a forward and backward moving photon. These two lines are also the asymptotes of the space- and timelike hyperbolas, because if w and x get very much larger than s, the curves approach the straight lines ever more closely.

What is the geometric meaning of these beautiful curves? What do they express? A good way to find out is by applying the Lorentz transformation formulas. If you take the formula for the hyperbola and for x and w you substitute the corresponding expressions in terms of w', x', and $\beta = v/c$ from the transformation rules on page 67, you will after some shuffling obtain the equation $w'^2 - x'^2 = s^2$. Which is exactly the same equation, but now in terms of the primed coordinates. This means that the curve for a fixed value of s is *invariant* under the Lorentz transformations! The transformations may move particular points back and forth on the curve, but the continuous set of points, the hyperbola as a whole, does not change.

In mathematics and physics we talk a lot about *vectors*. These are very much like arrows: they have a length and a direction. In Euclidean geometry – and therefore in ordinary space – the length r of a vector pointing from the origin to the point (x,y) squared is equal to the *sum* of the squares of its components, $r^2 = x^2 + y^2$, and the length is preserved under rotations. For a spacetime vector (w,x) we can define a similar quantity, namely the *spacetime interval s*, but its square is equal to the *difference* of the squares of the time and space components: $s^2 = w^2 - x^2$. The important point is that in relativity, it is the spacetime interval between two events that is preserved under the Lorentz transformations; it is the same for all inertial observers. Because of the minus sign in the definition, the square of the interval can be positive, negative or zero, in which cases we speak of a *timelike*, a *spacelike* or a *null* interval. Similarly if we draw an arrow between two events in spacetime, we speak of a timelike, spacelike or null vector. Indeed the vectors in the figure labeled x, x' and x'' are spacelike, while those labeled w, w' and w'' are timelike.

The timelike hyperbola (cutting through the x-axis like the time axis does) turns out to be interesting for another reason too. If you look at the timelike hyperbola, you see that it can in fact be interpreted as an entirely bona fide world line of some observer. That person is not traveling at a constant speed, but instead is continuously accelerating in the positive x direction. Though a full understanding of accelerated observers lies outside the scope of special relativity, we will nevertheless return to this particular observer towards the end of the book, exactly because we run into her world line here so naturally.

Circles and hyperbolas

In this section we further explore the hyperbola by comparing its properties with those of the circle. To do this, instead of considering Lorentz transformations we start out with ordinary rotations in the plane around the origin – which by definition leave a circle centered at the origin invariant. If we take a vector (say the red/blue arrow in the figure) of length r and rotate it, its endpoint traces a circle with radius r. In the figure for the red circle $r = 4$. Under rotations the endpoint of the vector moves around on the circle, but the circle as a whole remains unchanged. As we saw before, under the Lorenz transformations the head of the red/blue arrow will move along the blue hyperbola with $s = r$, and the latter too is invariant as a whole. For that reason the Lorentz transformations are sometimes called "hyperbolic rotations", as they leave the spacetime interval invariant.

In spite of the fact that we are working in a planar, two-dimensional world, there turns out to be something peculiar about the geometry underlying special relativity, originating in the minus sign between the time and space dependent terms in the definition of the invariant interval. I could indeed have chosen to confront you right from the start with a planar geometry where the "invariant length squared" of a vector is not defined as the sum of the squares of its components, but as the difference. This hyperbolic geometry is called Minkowski space, and that is the space we have tacitly been working in all along.

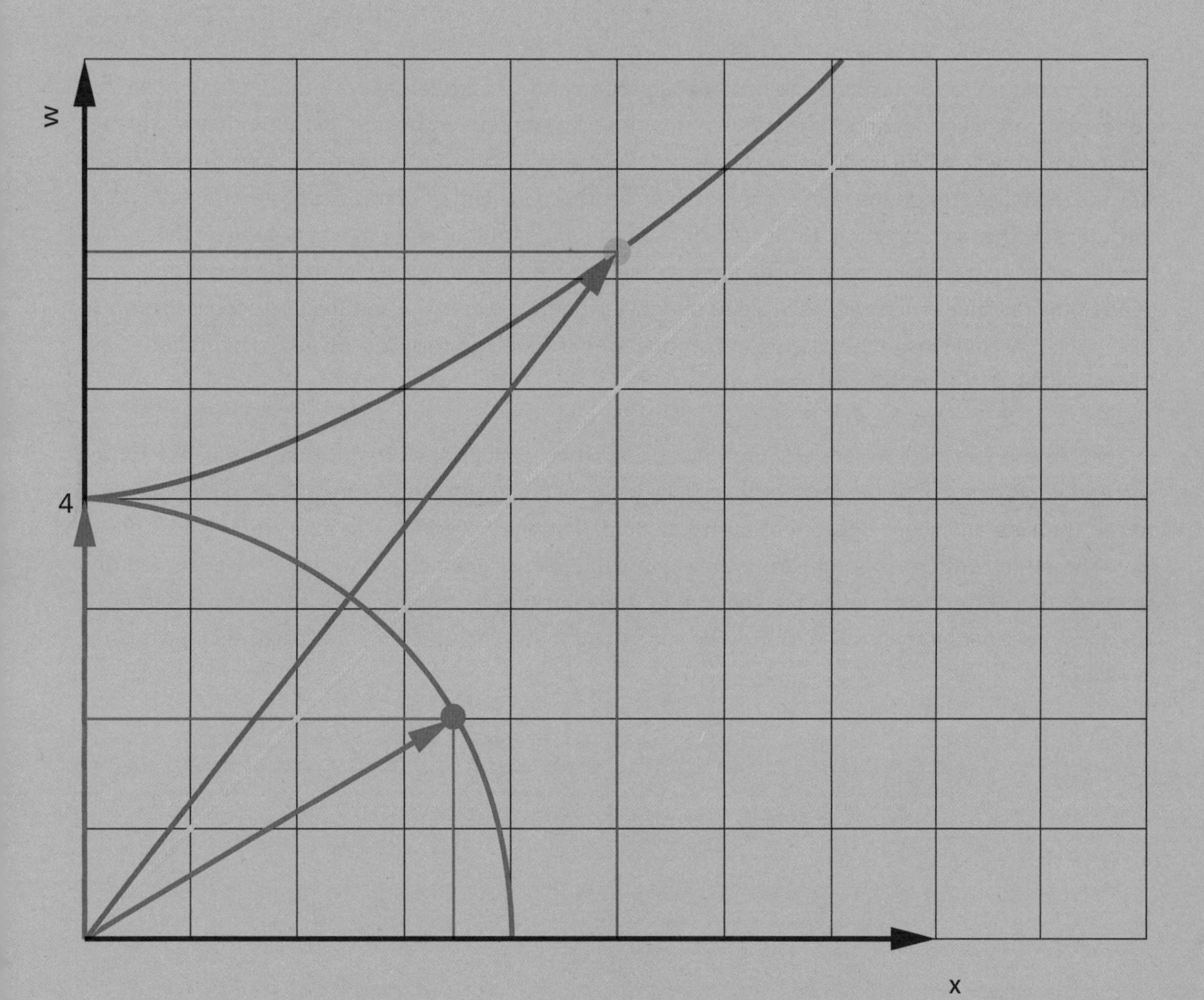
w
4
x

Constructing a hyperbola

It should be clear by now that hyperbolas are of great importance in relativity, being closely connected to the Lorentz transformations linking the inertial frames of different observers. This is why we shall explore the hyperbola a little more before we proceed with the physics of relativity.

You may have wondered why we are not getting rid of the minus sign by rewriting the equation $w^2 - x^2 = s^2$ in the form:

$$w^2 = s^2 + x^2$$

With this formula we can indeed conveniently construct the desired hyperbola, using a pair of compasses and the Pythagorean theorem. As you may recall, that theorem says that if s and x are the perpendicular edges of a rectangular triangle, then the long side of the triangle equals w, as defined by the equation above. But that is exactly the situation we are in: if in the figure we choose s as fixed along the vertical w-axis and we pick some point x along the horizontal axis, then the points x and s together with the origin define a rectangular triangle. The long side connecting x and s must then have the length w given by the formula above. If we now draw a circle centered at x with radius w, arching upward until it hits the vertical line through point x, we will have constructed a point (w,x) of the hyperbola. Different points of the hyperbola can thus be obtained from different points on the x-axis, as indicated in the figure. We see that one can indeed construct a hyperbola with ruler and compass, though it is admittedly more work than just drawing a circle.

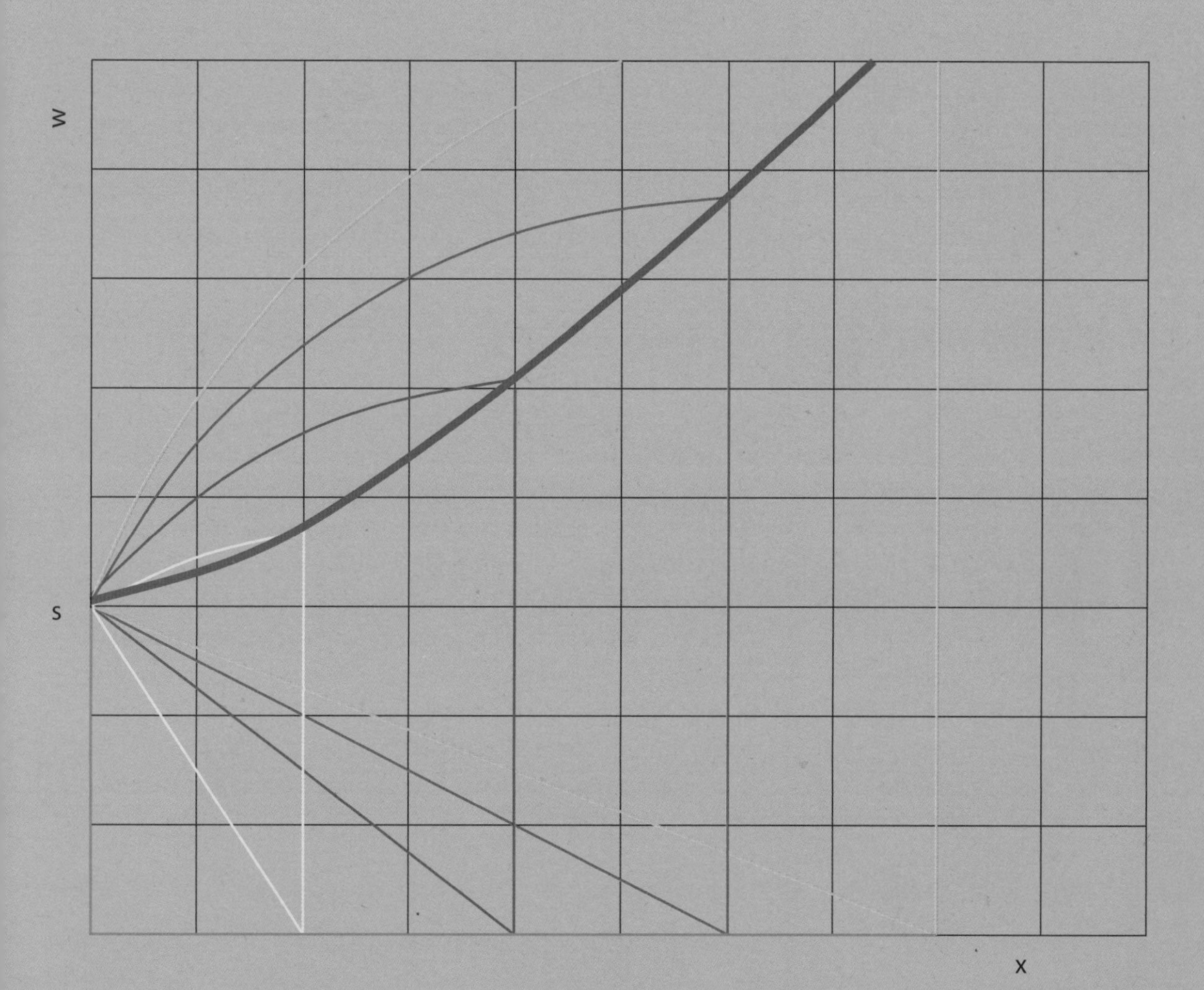
W
S
X

Wisdom on vectors

We have introduced the arrow-like vectors in ordinary Euclidean space and in spacetime. The normal space vectors, indicating a position or a velocity, have a length that is preserved under ordinary rotations. These rotations correspond with transformations from the frame of one stationary observer to the frame of another stationary observer, rotated with respect to the original one. In the context of relativity we are of course interested in spacetime vectors and their properties from the points of view of different inertial observers. We have distinguished two cases: in the non-relativistic case the coordinate frames and thus the vectors are related by the Galilean transformations, and in the relativistic case they are related by the Lorentz transformations. Furthermore the latter two are related, in the sense that the Lorentz transformations reduce to the Galileo ones for small values of β. We know that under the Lorentz transformations the time and space components of a vector transform in such a way that the endpoint of the resulting vector ends up on the same hyperbola. Both the direction and the length change, but the spacetime interval is invariant.

What does this mean? Let us start with the simple vector $(s,0)$ of length s pointing along the time axis, and transform it by the Lorentz transformations to the frame of an observer moving with a velocity for which $\beta = v/c$. By substituting $(w',x') = (s,0)$ into the transformation rules on page 67, we obtain the resulting vector $(w,x) = (\gamma s, \beta\gamma s) = \gamma s(1,\beta)$. So the final expression is a kind of spacetime velocity vector $(1,\beta)$, multiplied by a factor γs, which itself is velocity dependent (because γ depends on β). It looks rather complicated, but what is important is the fact that the resulting vector does satisfy the relation $w^2 - x^2 = s^2$, and therefore the transformation is in a sense simple. What I mean by "simple" in this context deserves some explanation.

Transformations such as rotations in the plane or Lorentz transformations in spacetime share the convenient property that the components of a vector, i.e. the w and x components, transform *linearly* into each other. The new x and w components are linear combinations of the old w' and x' component – and the converse is also true: it is a property of the transformation, not of a particular vector. If the transformation would involve, say, the squares or another function of the old components, then the transformation would not be linear. We have encountered a very nonlinear transformation already in the context of special relativity. Think of how the velocity parameter β_1 transforms when viewed from a frame moving with velocity parameter β_2; the resulting velocity factor is then given by Einstein's velocity addition formula as $\beta = (\beta_1 + \beta_2)/(1 + \beta_1\beta_2)$. I already pointed out on page 49 that the Einstein addition formula is nonlinear, in contrast to the Newtonian addition formula, which would just read $\beta = \beta_1 + \beta_2$.

Why make such a fuss about nonlinear versus linear? Surely nonlinearities make life more complicated, but as long as we have the formula, who cares? In our blessed days, we can after all ask a computer to perform the hairy algebra for us. It would be happy to do so! All of this being true, there is nevertheless an important physical reason why we really would like to somehow maintain linearity. This is why in the next chapter we return to physics, to discuss familiar concepts such as momentum and energy.

6 Energy and momentum

Once we accept our limits, we go beyond them.

A moving particle

We shall now discuss the notion of momentum for a moving particle, and in particular the differences between the classical Newtonian theory and special relativity.

In Newtonian mechanics the state of motion of a particle is characterized by its mass m and its velocity v or momentum $p=mv$. Newton came to the essential insight that a force causes a proportional acceleration a, where the proportionality constant is by definition the inertial mass m. His famous force law $F = ma$ basically states that force equals the change in momentum (per unit time). Velocity and momentum are vectors (as are force and acceleration); they have a direction and a length. In a three-dimensional world we think of vectors as arrows having three components, along the x-, y- and z-axes. In our toy world containing only one spatial dimension they can only point in the positive or negative x direction.

It should be clear by now that in relativity, space and time get mixed up in an essential way. This implies that we should not expect the conventional Newtonian velocity or momentum vector, which has only spatial components, to be adopted directly into relativity. We should look for a natural time component of the momentum vector, which allows us to define a spacetime momentum vector that transforms under either Lorentz or Galilean transformations like the spacetime position vector (x,t). To keep the discussion maximally transparent I will treat the two cases in parallel.

Our starting point is a particle at rest, and we ask ourselves what it looks like in a moving frame. The

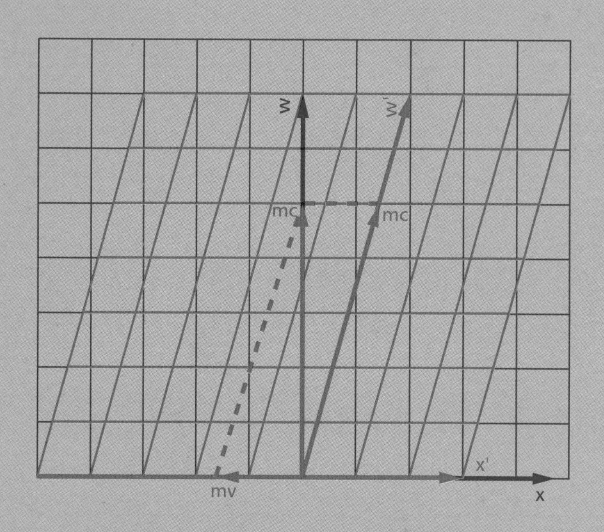
w
w'
mc
mc
x'
mv
x

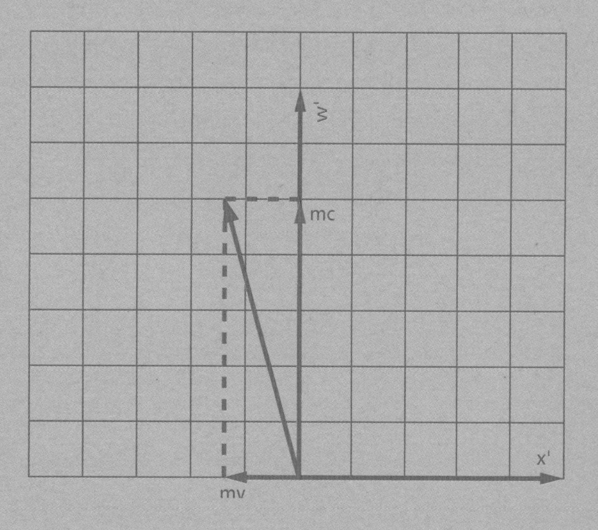
w'
mc
x'
mv

Newtonian (or Galilean) case is shown on the previous page. In the left figure the state of the particle has been depicted in terms of two parameters that characterize its motion (at a certain time): its mass and its momentum. Along the vertical axis we have put the quantity mc (a mass parameter), and along the horizontal axis the momentum $p = mv = \beta mc$. If we start with a particle of mass m at rest (so $p = 0$), its state is represented by the vector (arrow) along the vertical axis. We have also drawn a red frame in this same figure, corresponding to Newtonian observers moving with a velocity v. For them the particle moves with velocity $-v$, so, the momentum of the particle is $-mv$. It is crucial to observe that the picture of the frames is identical to the figures for the coordinates w and x. The reason is that they are related in exactly the same way through the Galilean transformation: for the spacetime position vector (w,x) we have $w' = w$ and $x' = x - vt = x - \beta w$, while for the spacetime momentum vector (mc,p) we have $(mc)' = mc$ (because the mass doesn't change) and $p' = p - mv = p - \beta mc = -\beta mc$. The vector representing the particle in the moving frame is shown in the right figure on page 85. To make the conceptual difference clear we now do exactly the same exercise for the relativistic case.

We start on the left again with the particle at rest; in the rest frame it is characterized by the vector whose time component equals mc and whose momentum is vanishing, $p = 0$. Now we would like to learn the magnitude of the components in the red frame. We could use the Lorentz transformation, or read it from the figure where we know that $(mc)' = \gamma mc$ and $p' = -\beta\gamma mc$, because of the rescaling of the red axes by the factor γ. The situation in the moving frame is represented in the figure on the right, where we see a relativistic spacetime momentum vector whose space component equals $-\beta\gamma mc$ and whose time component is γmc. This differs notably from the Newtonian result by the overall factor of γ, which implies that both the space and time components of the momentum vector go to infinity when the velocity of the moving frame approaches c. This also follows directly from the left figure: the components become increasingly parallel. At this point it is illuminating to reflect on the properties of light itself.

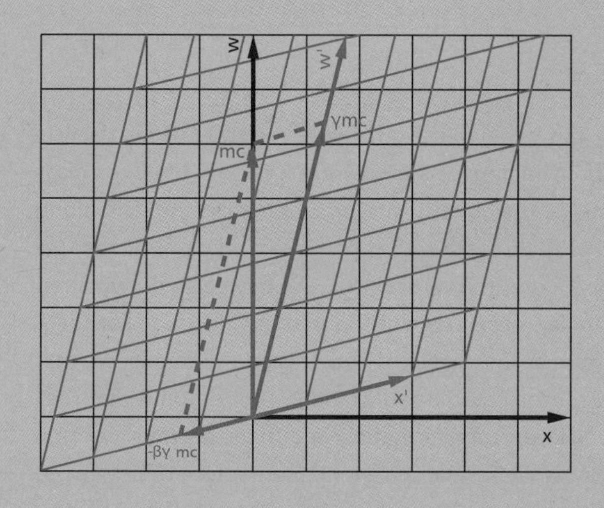
w
w'
γmc
mc
x'
x
-βγ mc

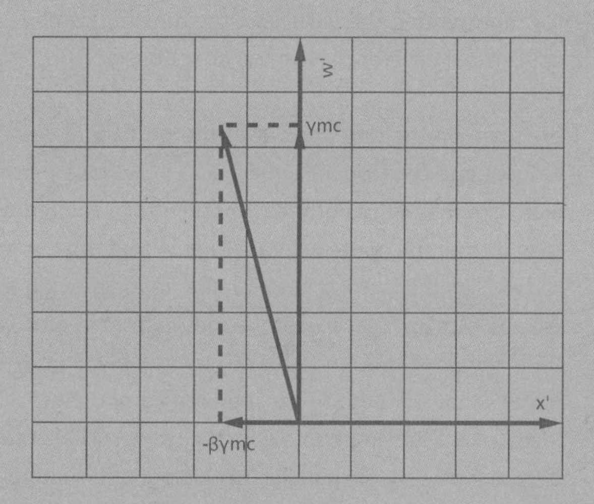
w'
γmc
x'
-βγmc

A photon or light particle by definition propagates with the speed of light, so its spacetime momentum is directed along the light-like world line and therefore has an equal space and time component for each observer. Maxwell's theory of electromagnetism tells us that light carries energy and momentum, and furthermore that their ratio is constant and universal. This ratio equals the velocity of light: $E/p = c$ for all observers. These remarks imply that the time component of the spacetime momentum vector should be identified with E/c. A photon travels at the speed of light, but even so it can have a finite momentum and energy, in contrast to the massive particles discussed above. In the figure the photon's energy-momentum vector is depicted in yellow. Note the unique feature that its components in all moving frames are located on the line perpendicular to the vector.

Contrary to a massive particle, whose momentum continues to increase without limit, light behaves nicely. The only way to understand the photon as a particle whose two components remain finite while γ is infinite, is to assign it a mass equal to zero, so we can replace the ill defined quantity γmc by the well defined energy E/c. This way, as a massless particle, the photon fits perfectly into the theory.

Brainteaser: Show that if the energy of a photon equals E in the rest frame, it equals $E' = (1 - \beta)\gamma E$ in a moving frame. Do this using the diagram on the right, and verify it by the applying a Lorentz transformation to the energy-momentum vector $(E/c,p) = (E/c,E/c)$. Compare the result also with that for the Doppler effect on page 58.

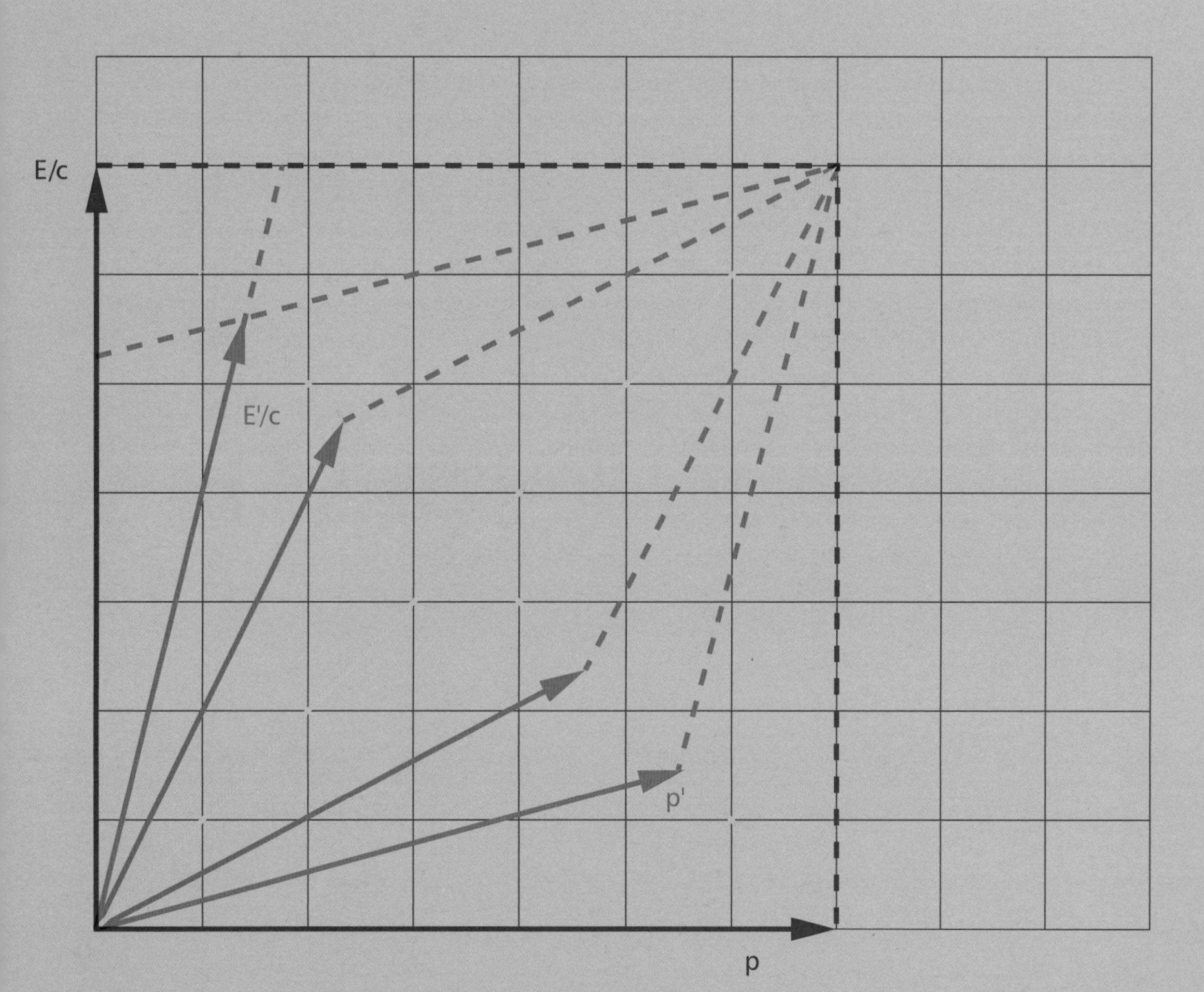
E/c
E'/c
p'
p

In the previous section we introduced a "spacetime" momentum vector (γmc, $\beta\gamma mc$) for massive particles, which followed quite naturally from the relativistic relations between different frames. The perspective also accommodates photons, by assigning them an energy-momentum vector *($E/c, p$)*, where $E/c = \pm p$.

The massive particles' space component $\beta\gamma mc$ should be interpreted as their physical momentum p, because in the limit $\beta = v/c \rightarrow 0$ we have $\gamma \rightarrow 1$, and thus $(\gamma mc, \beta\gamma mc) \rightarrow (mc, \beta mc)$. In this perspective the quantity γm gets the natural physical interpretation as the relativistic generalization of the mass m, and that is exactly what Einstein proposed. He defined the relativistic mass as $m_{rel} = \gamma m$ (which as you can see depends on the velocity). Comparing the time component to the photon's, we arrive at the startling conclusion which Einstein had to draw in 1905, namely that $E = m_{rel}c^2$. This is the famous equation expressing the equivalence of energy and mass, celebrated because of its unequalled simplicity, power and beauty.

To understand why the time component of the energy momentum vector corresponds to the relativistic energy of a massive particle, it is illuminating to look at the expression at small velocities. We may then approximate the expression for γ, assuming that β is very small, to obtain:

$$m\gamma = \frac{m}{\sqrt{1-\beta^2}} \cong m + \frac{1}{2}m\beta^2 + \cdots$$

The first term on the right hand side is just the mass, as expected. The second term contains β^2, and the dots represent terms containing higher powers of β, which are negligible because we assume β to be

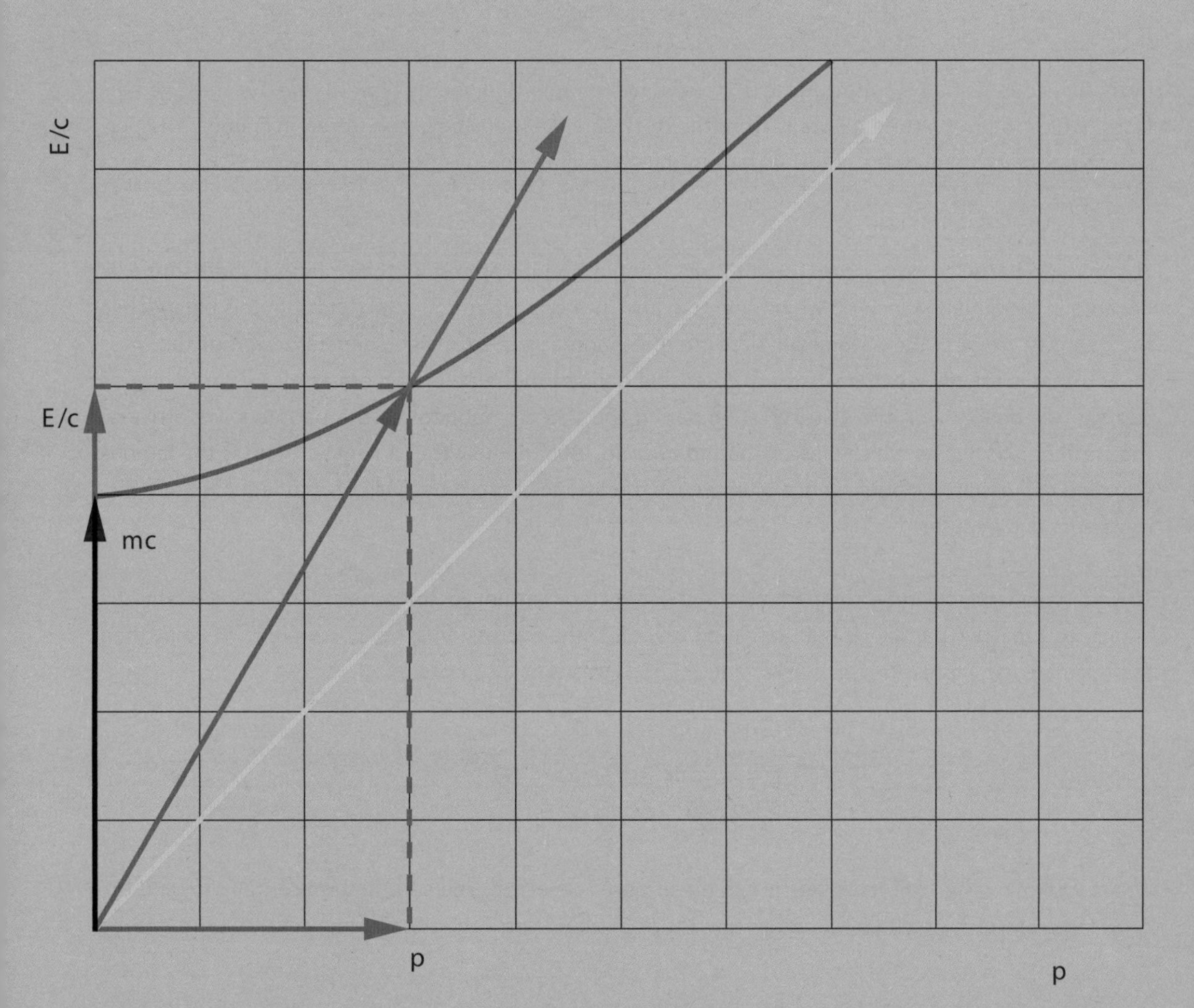
E/c
E/c
mc
p
p

small. The second term can also be given as $\frac{1}{2} mv^2/c^2$, which is (up to a factor c^2) exactly the expression for the kinetic energy of a particle of mass m and velocity v in the Newtonian theory. So we find that the relativistic mass of a massive particle at reasonably low speed can be approximated by its Newtonian mass plus its kinetic energy divided by c^2. The time component of the relativistic momentum vector is entwined indeed with the particle's energy; hence the term *energy-momentum vector.*

I find it quite exciting to see how elementary reasoning, consistently carried through, can lead to such a revolutionary insight as 'mass is just a particular form of energy'. One gram of any kind of matter corresponds roughly to 10^{17} Joules, comparable to the energy released during the bombing of Hiroshima.

Nowadays physicists actually prefer a slightly different terminology when talking about the formula above: they speak about the invariant or rest mass m corresponding to the invariant length of the relativistic momentum vector (E,pc). As a formula: $E^2 - p^2c^2 = m^2c^4$. This expression applies to photons and other massless particles as well: if we set $m = 0$, it correctly yields $E = \pm pc$.
The figure on the previous page elegantly summarizes all features of energy and momentum as they appear in different frames. Note that the picture is very similar to the one for the relativistic position vector (w,x) on page 79.

In the following chapter we will look at the cherished conservation laws for energy and momentum and consider systems of two colliding particles. If you prefer, you can also skip ahead to chapter 8 where we discuss accelerated observers.

Fusion and fission

The equivalence of mass and energy, so concisely expressed by the equation $E = m_{rel}c^2$, has tremendous applications. These are most dramatic in the realm of nuclear fusion and fission. Nuclei are tightly bound systems composed of a total number of protons and neutrons. Because these "nucleons" are held together by the strong force, each nucleus has a characteristic binding energy per nucleon. The figure shows the binding energy per nucleon as a function of the total number of nucleons or the atomic number N. On the left we see that for small atomic numbers we may lower the binding energy per nucleon by fusing simple nuclei into more stable composites, such as in the reaction: $D + T \rightarrow {}^4He + n$ + energy. The difference in binding energy is released, and the amount is typically a million times larger than in an elementary chemical reaction. On the other side of the mass scale we find heavy nuclei like Uranium, which may be metastable and decay into nuclei of lower mass, thereby also producing extra energy. This process of fission is the working principle of our present nuclear reactors. In the long term fusion reactors are expected to become technically feasible, which would be a preferable option from the point of view of safety and radioactive waste management. As fuel for fusion is cheaply available in virtually unlimited amounts, this may be the ultimate solution for global energy needs in the long run.

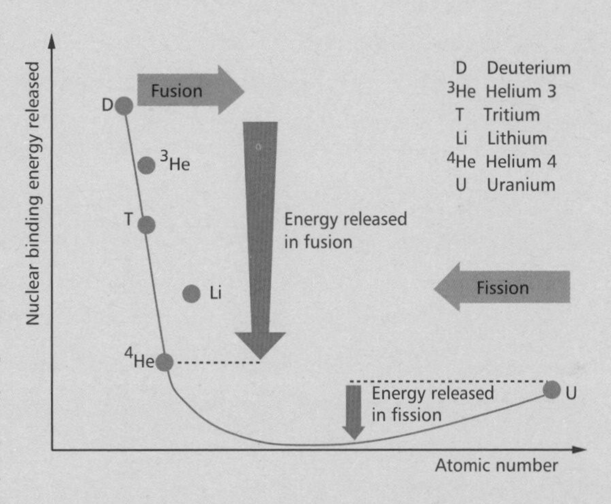

7 The conservation laws

The law of conservation of mass lost its sovereignty and was absorbed into the law of energy conservation.

Total momentum

We speak of a conservation law if in some process certain quantities do not change. In an electric circuit, charge moves around freely, but cannot get lost. If we burn something, than the law of Lavoisier tells us that the total mass in a closed system will not change. In a building lots of people may move around, yet the total number of people in the building will not change (except of course if they are in a birth clinic). You see that there are all kinds of conservation laws. Here we focus on what happens to the conservation laws for mass, momentum and energy which hold in Newtonian particle dynamics, if we look at them from a relativistic point of view.

To understand the conservation of momentum in Newtonian mechanics, we first look at the simplest case: a single particle on which no force is working. When we apply Newton's second law $F = ma$ to this system and set $F = 0$, mass times acceleration vanishes: $ma = 0$. As force equals the change of momentum per unit time, we conclude that when no force is applied to a particle, its momentum is conserved. The next step is to consider systems consisting of two colliding particles. Although the particles will exert a force on each during the collision, there is no external force. Since no external force is applied to the system as a whole, the total momentum – which is simply the sum of the individual particle momenta – is conserved.

We start by graphically representing the situation before the collision. In the figure on the previous page we have drawn the incoming momentum vectors of the type we introduced in the previous section, and also their sum, the total incoming momentum vector p. Its time component is given by the sum of the masses, which amounts to the total mass of the system. In the sort of (one-dimensional)

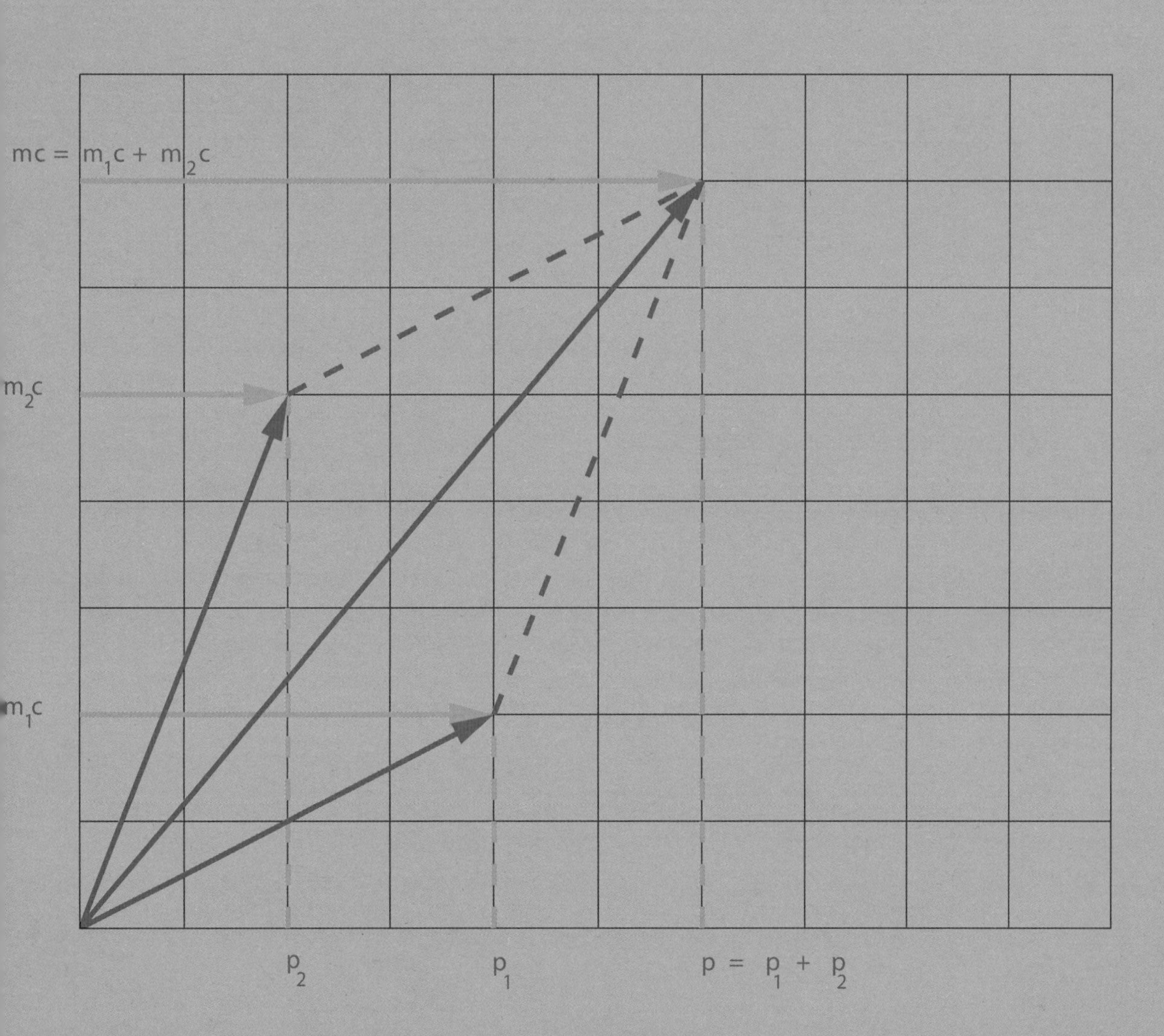

mc = m1c + m2c
m2c
m1c
p2
p1
p = p1 + p2

billiard we are talking about here, mass is a conserved quantity. Note that particle 1 has the largest velocity, so it better be on the left of particle 2: otherwise no collision will occur.

The total momentum vector has a direct physical interpretation: it represents the situation where the colliding particles stick together and move forward as a single particle with mass $m = m_1 + m_2$ and momentum p. This is called a completely inelastic collision, because momentum and mass are conserved but (kinetic) energy is not, as will become clear shortly.

Because we already know what a relativistic energy-momentum vector looks like from the previous chapter, it is not hard to generalize the figure to the relativistic case. We replace the mass component by an energy component that is dependent on the momentum, so that we obtain the characteristic hyperbolic curves on which the momenta live.

Observe that the time component of the total momentum is still the sum of the two individual time components: the total energy is just the sum of the energies of the individual particles. However, the invariant mass associated with the total momentum, given by the point where the upper hyperbola for the total momentum intersects the energy axis, is not equal to the sum of the rest masses of the individual particles – It is in fact larger! A possible example of this is the decay of a particle into two lighter particles. Total energy is conserved but mass is not: It is partially converted to kinetic energy.

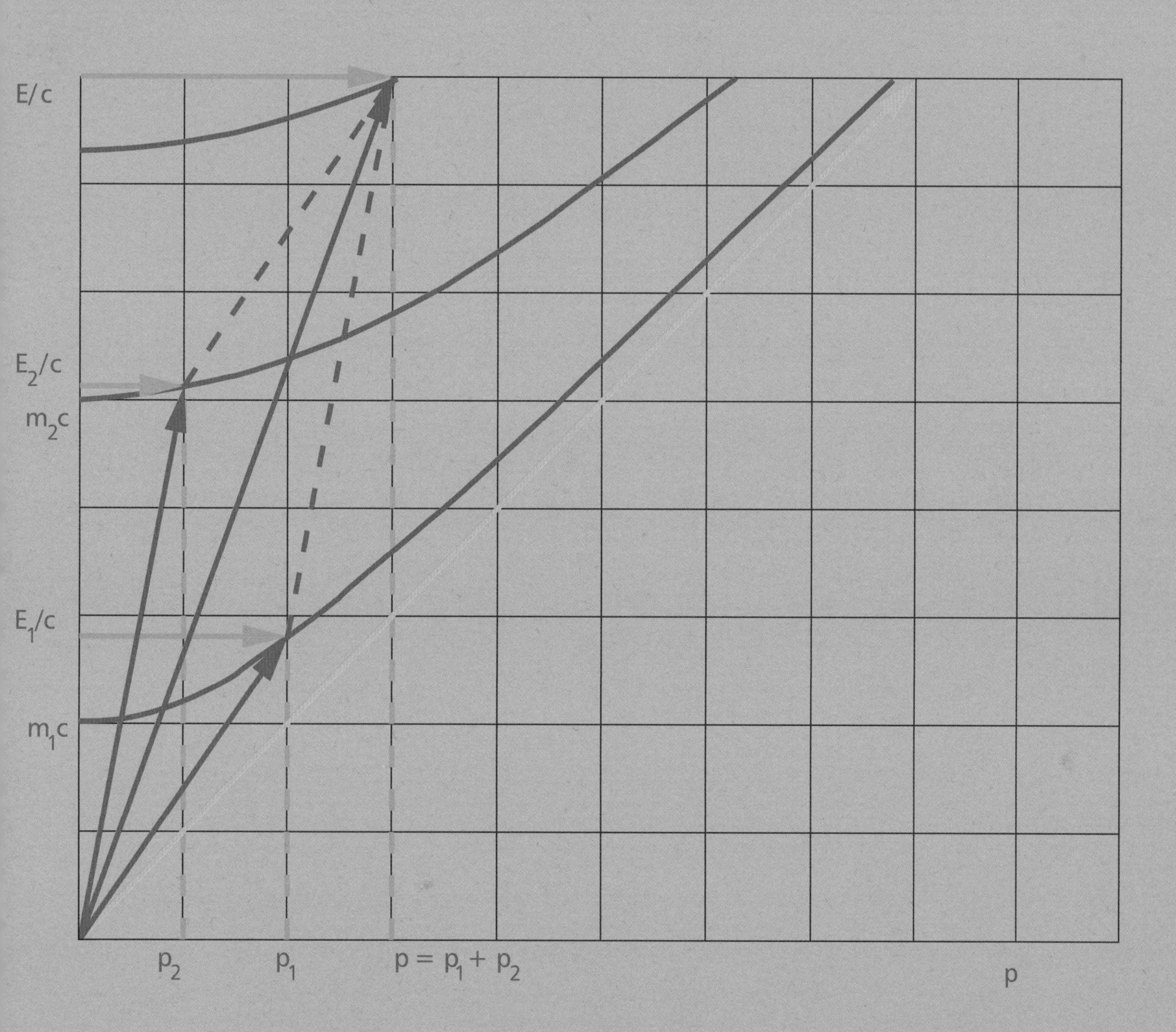

E/c
E_2/c
m_2c
E_1/c
m_1c
p_2
p_1
$p = p_1 + p_2$
p

Momentum in a moving frame

The figures we presented on page 85, with the mass parameter mc as the time component of momentum, are not standard in any treatment of classical mechanics, but they are extremely useful in visualizing the basic discrepancy between classical and relativistic mechanics. Before getting into a discussion of momentum conservation let us see what the situation of two colliding particles looks like in a moving frame, say a frame that moves with a velocity u. For the Newtonian version, all we have to do is apply the Galilean transformation to all velocities and see how the momenta change. The particles' velocities will change according to $v' = v - u$, and thus $p' = p - mu$. The figure shows the effect from the point of view of a moving (Newtonian) observer. We see that the changes in the momenta can be read off very easily. The red line for the moving observer is the same as it would be in the spacetime diagram. Referring to the light blue arrows, it is obvious from the figure that also in the moving frame the total momentum is just the sum of the two particle momenta. After all, the change of frame has not affected their mass and the arrows still add up correctly.

The freedom to analyze the situation in any frame allows us to pick a particularly convenient frame to do so. One such frame is the so-called "zero momentum frame": the frame in which the space component of the system's total momentum is zero. This is the frame in which the red line coincides with the total momentum arrow. To see what the situation looks like in that frame, we move to the next figure.

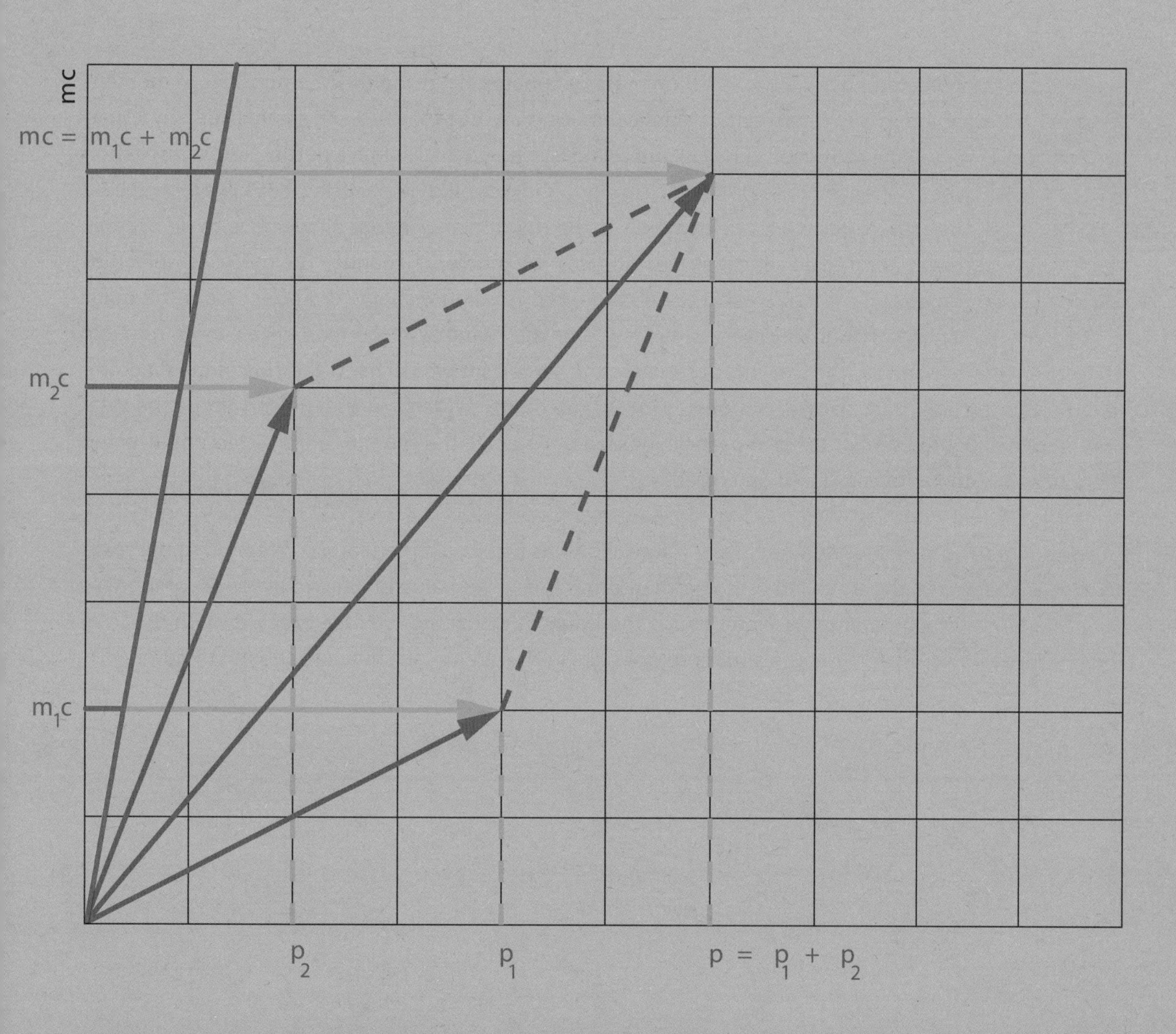
mc
mc = m1c + m2c
m2c
m1c
p2
p1
p = p1 + p2

In this figure we see both the incoming and the outgoing momentum vectors (the solid arrows p_1 and p_2, and the dashed arrows P_1 and P_2 respectively) for the collision experiment discussed before. In the zero momentum frame the total incoming spatial momentum p is zero, and its time component equals the sum of the masses. Observe that in this frame one particle moves to the right and the other one to the left. Momentum conservation is the statement that the total momentum vector must be the same before and after the collision. Its spatial component thus remains $p = P = 0$, while the vertical component is again just the sum of the masses $mc = m_1c + m_2c$. The restriction of momentum conservation translates into the requirement that the horizontal components must be equal and opposite, both before and after the collision, and we have drawn one particular instance of that. If we combine this picture with the previous one, it is evident that if momentum conservation holds in one (Newtonian) frame, it also holds in any other. Switching frames only amounts to drawing some red line representing the relative velocity.

Note that the conservation of momentum does not fix the outgoing momenta of the individual particles completely, but only their sum. In our one-dimensional situation, one more relation would suffice to fix them completely – such as an energy condition, and we will return to this option shortly.

Now, you may wonder why we are spending so much effort on this issue of momentum conservation. The answer is that we want to understand what will happen to these very simple pictures, in which we can easily move from one frame to another, if we consider the situation not from the Newtonian perspective, but from Einstein's.

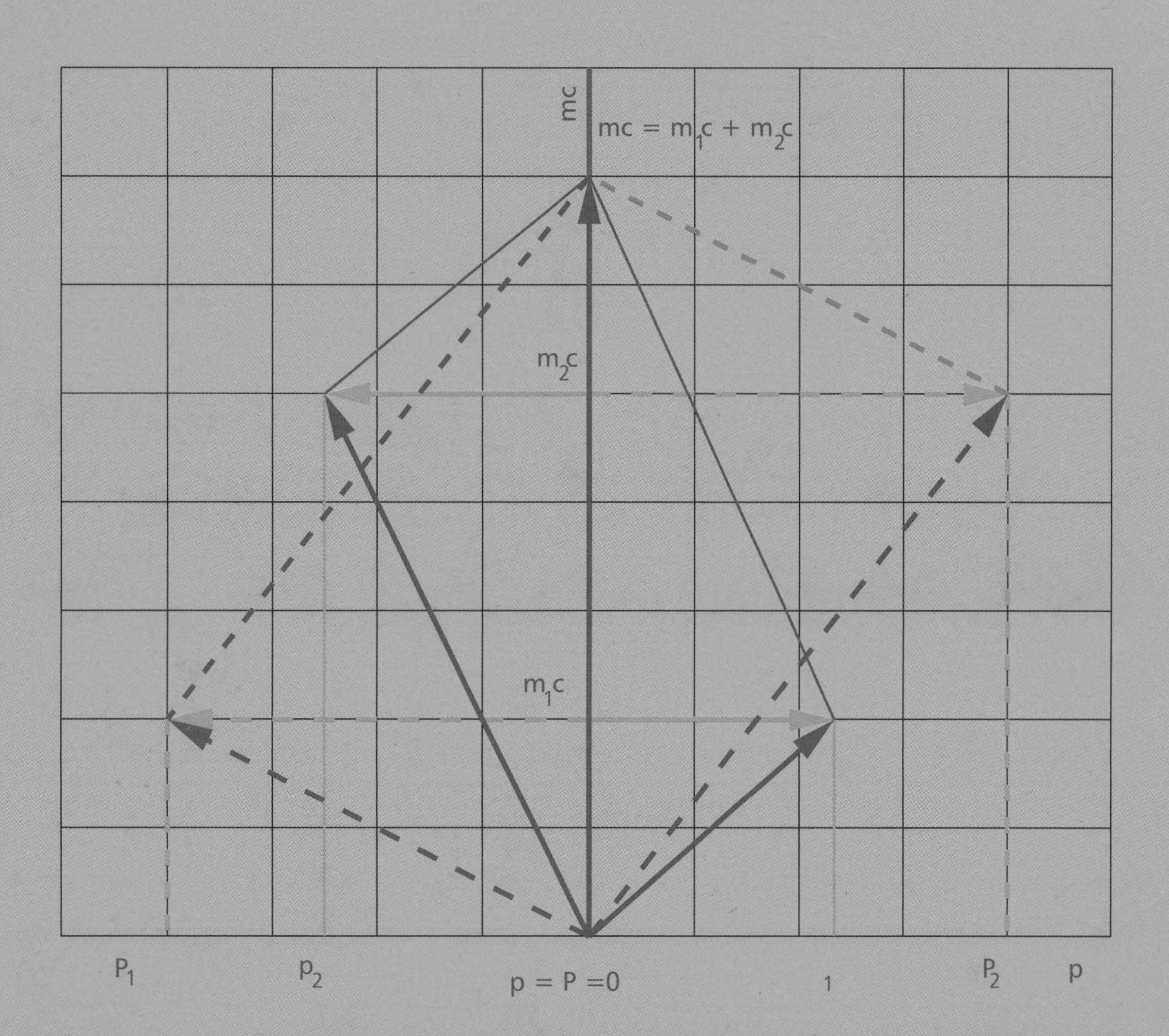
mc
mc = m1c + m2c
m2c
m1c
P1
p2
p = P =0
1
P2
p

What happens if we replace the Galilean transformation rule by its Lorentzian counterpart? Well, then the whole thing falls apart, because we have to replace the simple transformation of velocities $v' = v - u$ by Einstein's formula $v' = (v - u)/(1 - vu/c^2)$. This transformation is very nonlinear, causing the sum of the momenta before and after the transformation to differ. If we were to take Newton's definition of momentum and then apply the Lorentz transformations to it, we would come to the atrocious conclusion that the conservation of momentum no longer holds for all observers. So Einstein was faced with a simple choice: either give up the sacred law of momentum conservation, or come up with a different definition of momentum. As we saw in the previous chapter, he chose the second option. It was a choice with dramatic implications, which have however been substantiated experimentally. Having brought together the notions of space and time, he now had to do the same for energy and momentum.

Another very basic conservation law, cherished in all areas of physics, is the conservation of energy. Let us briefly recall what it means in the Newtonian framework. If we look at the collision of two billiard balls, the collision will be (almost) "elastic", meaning that the total kinetic energy (motion energy) of both balls together is strictly conserved. The kinetic energy E of an object is quadratic in the velocity (or the momentum), namely for the first ball $E_1 = ½\, m_1 v_1^2 = p_1^2/2m$. Plotting E_1 as a function of p_1 gives the dashed parabolic curve as shown in the figure.
If instead we allow two balls of clay to collide, clearly they would stick together and after the collision they would both continue with the same velocity: $v'_1 = v'_2$. In the zero momentum frame the resulting velocity is zero and all kinetic energy gets lost. Such a collision is called completely inelastic. It is not hard to imagine many cases in between these two extremes. Actually we know that in an inelastic collision the energy does not really get lost: it is converted into *interna*l motion of the molecules in the ball, which heats up or gets permanently deformed, changing its internal energy.

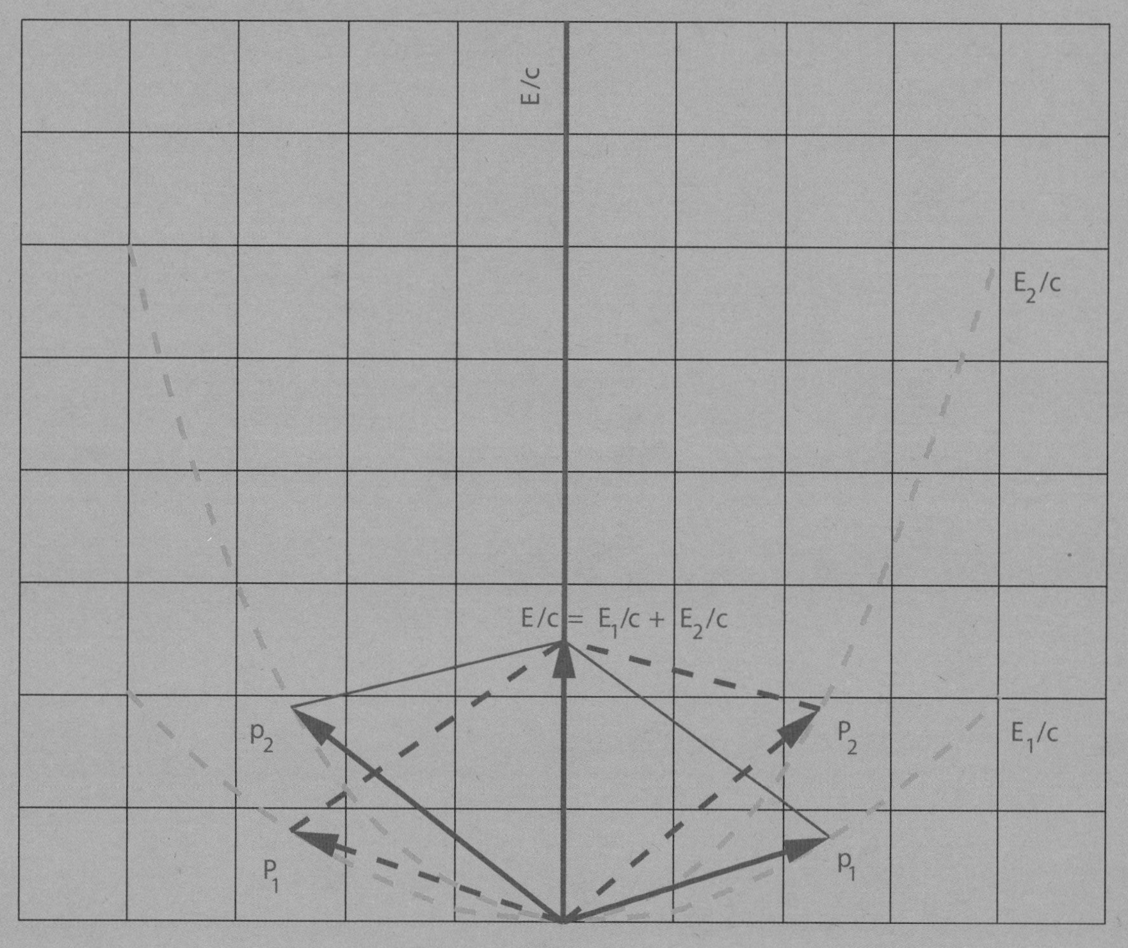
E/c
E_2/c
$E/c = E_1/c + E_2/c$
p_2
P_2
E_1/c
P_1
p_1
p

In the figure on the previous page we have put the momentum of a particle along the horizontal axis, and the corresponding energy along the vertical axis. You can see how the various energies (E_1 and E_2) depend on the respective momenta in the zero total momentum frame. (The total momentum vector points along the time axis and $p_1 = -p_2$.) The energy-momentum vectors of the two incoming particles are the solid arrows; those of the outgoing particles, the dashed arrows. The total kinetic energy according to Newton is obtained by adding the corresponding energies of the individual particles, or for that matter by adding the incoming or outgoing vectors. In the figure we show a completely elastic collision, for which the total energy before and after the collision is the same. It is immediately clear from the figure that (in one spatial dimension) the *only* solution is now the situation where the particles exchange momenta, so, $P_1 = p_2 = -p_1$ and $P_2 = p_1 = -p_2$. That is why the two E_1 curves in the figure coincide, just like those E_2.

The conservation law for energy and momentum is the relativistic equivalent of the classical conservation laws for momentum, mass and energy. Note that mass apparently is no longer conserved independently, because it forms part of the total relativistic energy. We arrive at the single relativistic figure on the facing page. You can see that for small momenta it can be approximated by "adding" the previous two non-relativistic diagrams of pages 101 and 103. A synthesis of sublime simplicity.

Brainteaser: A *pion* is a particle with a mass of 273 times the electron mass. It is unstable: it decays into a *muon*, with a mass of 207 times the electron mass, and an *antineutrino*, with a negligible mass. Sketch the energy-momentum diagram for this decay reaction in the rest frame of the pion. Algebra lovers should also use the conservation of energy and momentum to calculate the energies of the two decay products.

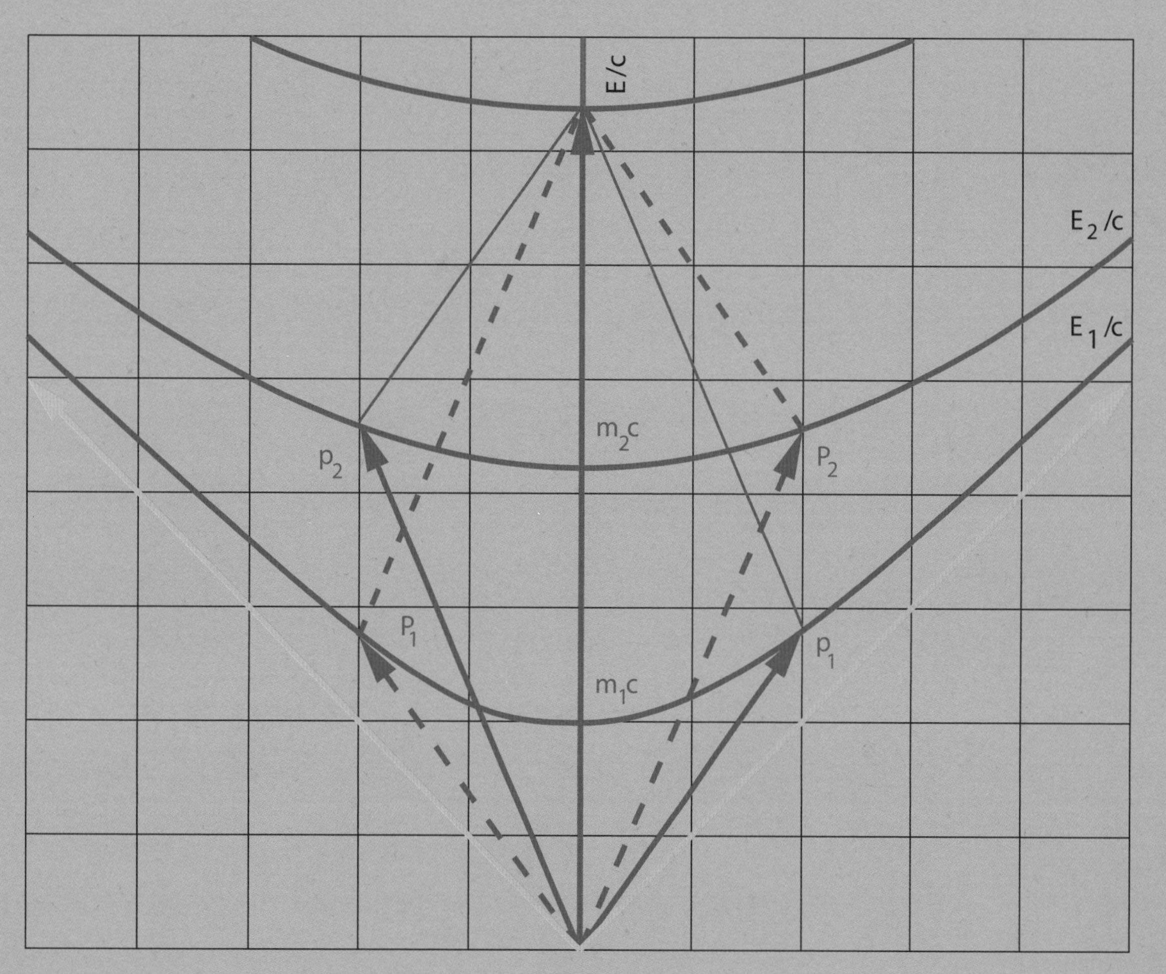
E/c
E_2/c
E_1/c
m_2c
p_2
P_2
P_1
p_1
m_1c
p

Large colliders

There are a few places in the world where relativity is bread and butter business. In the US, Europe and Japan large accelerators have been built in order to give elementary particles extremely high energies, meaning their speed approaches the velocity of light. A common design for an accelerator is to have a large circular ring in which the particles are accelerated and "stored". Usually there are two beams running in opposite directions, consisting of particles that have the same mass and often opposite charge. Because the momenta of the particles in the two beams are equal and opposite, the lab is exactly in the zero momentum frame. The particles are made to collide head-on in certain interaction regions, releasing enormous amounts of energy, which can then be converted into new types of matter such as very heavy particles. For instance, physicists are hoping to encounter the so far hypothetical Higgs particle.

In 2007/2008 the largest accelerator ever built, the Large Hadron Collider (LHC), will come in operation at the accelerator laboratory CERN in Geneva, Switzerland. It has a circumference of 27 kilometers and will accelerate protons to energies that are equivalent to about 7,000 times their mass, so that $E = \gamma mc^2 = 7{,}000\ mc^2$. Thus we have $\gamma = 7 \times 10^3$, and using the definition of γ one easily calculates that β is about $1 - 10^{-8} = 0.99999999$. So the velocity of these protons is amazingly close to the speed of light. This is very much out of scale with the pictures we have been drawing so far. In the figure on the previous page there would be only one hyperbola for a pair of colliding protons, as they have huge but opposite momenta. The point E would be 14,000 times higher than the intersection of the hyperbola with the vertical axis! An extremely relativistic situation.

Tachyons

In chapter 3 we argued that particles could not move faster than the speed of light, which saved the sacred notion of causality. Still one may ask whether it is admissible to *a priori* introduce particles that move faster than light. Such hypothetical particles are called *tachyons*, and what relativity has to say about them can be readily understood using a spacetime diagram. From relativity it follows that the tachyon is a particle with a spacelike energy-momentum vector having negative mass squared $m^2 = -\mu^2$. Its invariant energy-momentum relation reads $E^2+\mu^2c^4=p^2c^2$, and the corresponding curve in the (E,pc) plane is a timelike hyperbola (crossing the x- or p-axis) like the one depicted in the figure on page 75. Observe that for all points on that curve the momentum p is larger or equal to μc, in other words v is never smaller than c. For other inertial observers the energy momentum vector of the tachyon will move over the entire hyperbola and one is forced to accept negative energy states. The conclusion is that the existence of tachyons is not excluded by relativity, but if they can interact with ordinary matter, the fact that they move faster than light allows for violations of causality, while their negative energy states will cause instability of matter. It is comforting to know that so far they have only been encountered in science-fiction novels.

Brainteaser: Assume that tachyons exist and do interact with ordinary particles. Show in a diagram similar to the one on page 105, combining the energy-momentum vectors of a spacelike and a timelike hyperbola, that the conservation of total energy and momentum would allow for processes where an ordinary particle could emit a tachyon. Check that such processes are not allowed with the emission of an ordinary particle or photon. This signals that tachyons can cause matter instability – a significant argument against their existence.

8 Beyond special relativity

The important thing is to not stop questioning.

Tensions

Having mastered some special relativity, we now are ready for yet another challenge, where the velocity of observers will not remain constant. Imagine two rockets, Apollo and Sputnik, flying one behind the other. They are moving with the same velocity, so they are at rest with respect to each other. Between them there is a tight inelastic rope. The pilots have agreed that on a predetermined time they both will accelerate at exactly the same rate. As they intend to do that at exactly the same time, you might think nothing will happen to the rope. However, contrary to what you might expect, something surprising does happen.

Again it is profitable to analyze the situation in a spacetime diagram. The figure depicts the situation in a somewhat idealized form. First we see A and S at rest (with respect to a conveniently chosen frame of reference); then at time w_0 they both change velocity. Viewed from the moving red frame, the distance between the two rockets first corresponds to the double arrow labeled 1, which is also the length of the rope as viewed from this frame. Then S takes off first, and the distance increases to the longer double arrow labeled 2. The rope however –being inelastic – keeps the same length 1 in its rest frame (the red frame), and will therefore break. For the black observers at rest both rockets do change velocity at the same time and the distance between them remains constant indeed (see the black arrows in the figure), but for them the rope is Lorentz-contracted once it moves. A moving rope of length 1 takes on the contracted length of the dashed black arrow in the rest frame, which for the observers at rest explains why the rope will break.

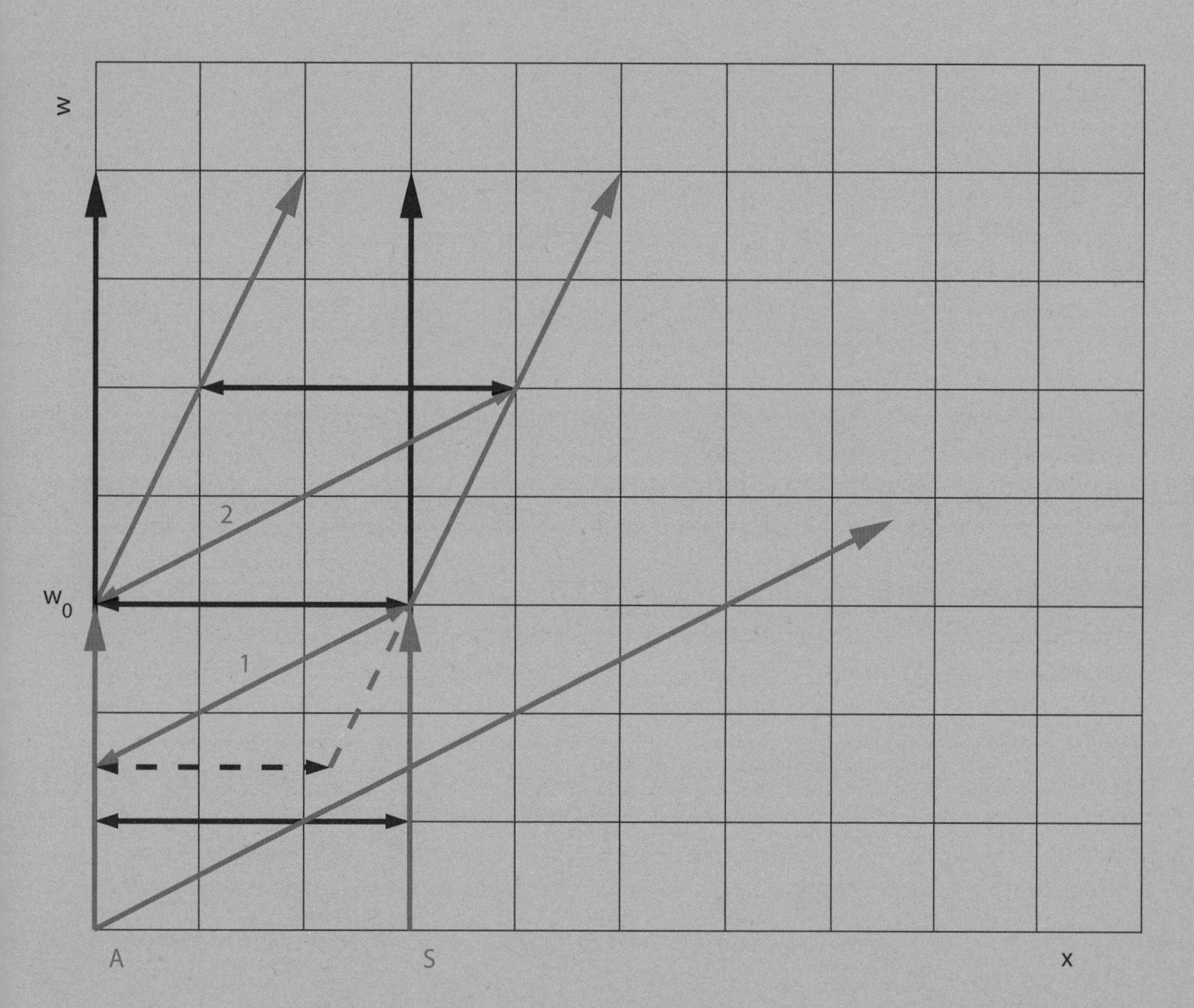
w
2
w0
1
A
S
x

An accelerated observer with horizon

We are approaching the almost-happy end of our story. I should tell you that after the papers on special relativity, Einstein kept quiet for quite some years and then came back with another theory that shocked the world, called *general relativity*. In that theory he showed how his idea of relativity could be generalized to situations where observers move arbitrarily with respect to one another. Stated differently, it extended the notion of Lorentz invariance to an invariance under arbitrary coordinate transformations. This theory turned out to be a new theory of gravity and is considered one of the greatest achievements in all of science.

Here we will only touch on some elementary aspects of it superficially, by considering an accelerated observer. And not any observer: We choose a particular example, namely the traveler whose world line is the red hyperbola in the figure. This is in fact nothing but the timelike hyperbola we discussed on page 75. Clearly the forward velocity of this traveler is increasing steadily. We see that her velocity for large times tends to the velocity of light, while the acceleration keeps decreasing. At any given point on her world line her frame is formed by the ray from the origin through her position as the space axis, and the tangent to the hyperbola at her position as the time axis (not drawn, except in the origin). Therefore the velocity parameter $\beta = v/c$ will depend on time: $\beta = \beta(w)$. Since the velocity parameter is just the tangent of the angle between the moving frame axes and the corresponding axes of the rest frame, we immediately conclude that at each point (w,x) on the world line of our observer we have $\beta = w/x$. Combining this with the defining equation of the timelike hyperbola, $x^2 - w^2 = s^2$, we obtain immediately that $\beta(w) = w/\sqrt{(w^2 + s^2)}$, which indeed tends to unity if w approaches infinity. We also derive that the scale factor γ equals $\gamma = 1/\sqrt{(1 - \beta^2)} = (1/s)\sqrt{(w^2 + s^2)}$.

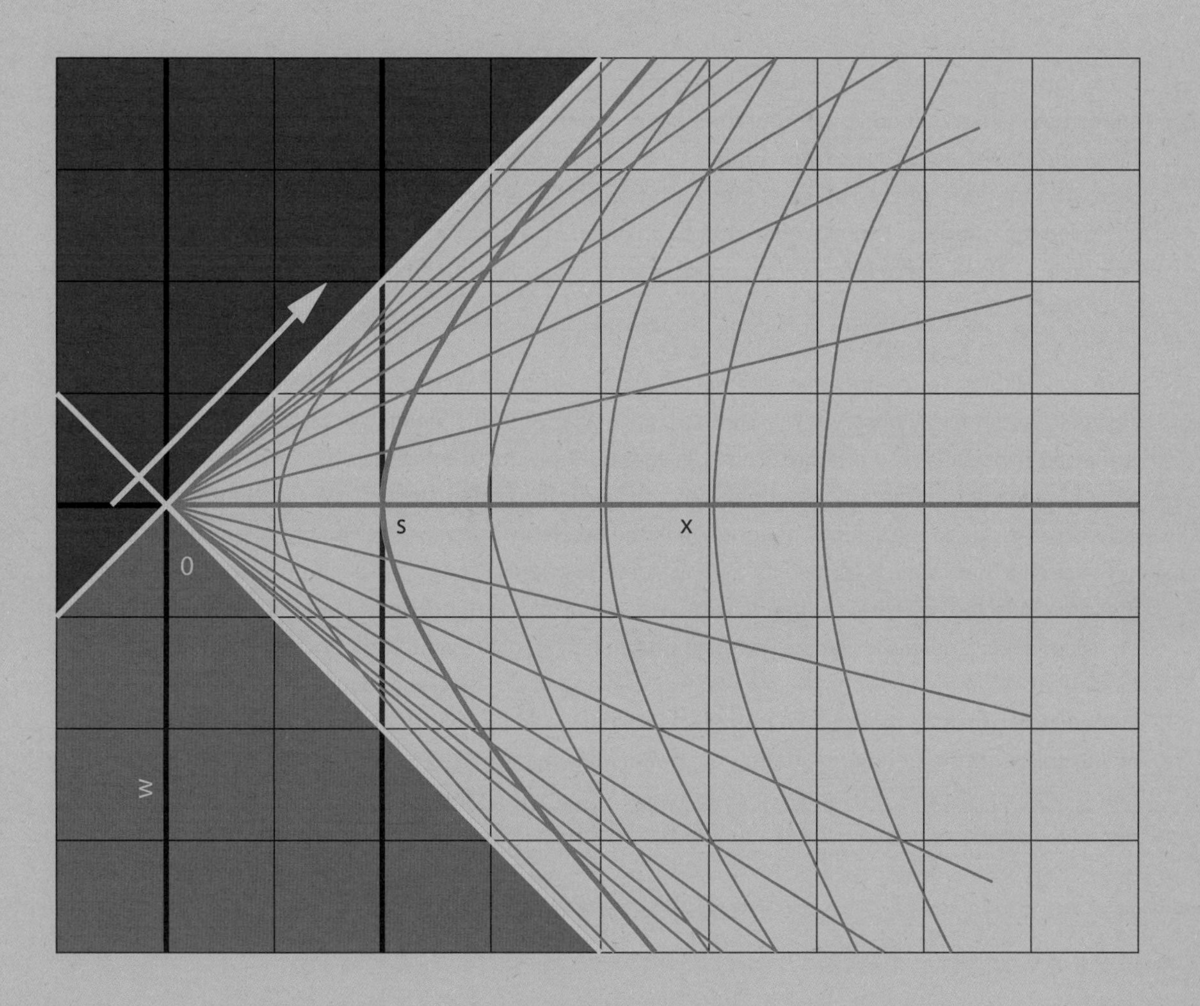
S
X
0
W

Clearly our observer is in a special situation. What is so special about it is most easily seen if we look at the relativistic force law, $F = dp/dt = dpc/dw = d(\beta\gamma mc^2)/dw$ where p is the relativistic momentum. In this case the calculation becomes utterly trivial because after substituting the values for β and γ, we obtain that the momentum is $\beta\gamma mc = wmc/s$, which is just w multiplied by a fixed number. The fact that the momentum increases linearly with w means that the force F is constant: $F = mc^2/s$, which does not depend on w at all. The conclusion is elegant and simple: a timelike hyperbola corresponds to the world line of an observer on whom a constant force is exerted! The parameter s characterizing the hyperbola equals $s = mc^2/F$ and is basically fixed by the force to mass ratio.

A typical physical realization of the situation we are describing is a charged particle traveling in a constant electric field (in the x direction). In relativity the constant force does not lead to a constant acceleration, because the mass is no longer constant but increases with velocity, so that the velocity can never exceed the speed of light.

How will the observer, who is by definition at rest in her own frame, experience this constant force? For her the situation is very similar to standing in an elevator, which is accelerating, and she will interpret the force as a kind of gravitational force, because accelerating upward makes you feel heavier. Here we catch a glimpse of the general theory of relativity, based on the equivalence of accelerated observers and the effect of a gravitational field, or in other words between inertial and gravitational mass.

We have found a physical interpretation for the timelike hyperbola by exploiting our knowledge of geometry which we built up step by step in the previous sections. Now we should look at the figure once more, because there is yet another surprise in store. What happens to all observers who have stayed behind in the rest frame, corresponding to strictly vertical world lines? We can imagine them exchanging messages with our traveler in order to keep her posted on the situation back home. Let us say that they have wireless communication using signals that travel with the speed of light. What do we infer from the diagram? We see that the signals from a given stationary observer (say, the one moving straight upward along the black line $x = s$) will reach her without problem until the stationary observer "enters" the dark region (at $w = s$). From that moment onward his messages will no longer reach her, even after an arbitrarily long time. The future light cones of events in the dark region lie entirely within the dark region. On the other hand, the messages that the traveler sends to him will reach him without problem at all times. This drastic phenomenon is typical for situations involving accelerated observers. Spacetime gets divided up into distinct regions, which are separated by an *event horizon*. The traveler outside it will never know about the events that take place behind this horizon, as for her those events stay hidden in the dark forever. The figure also shows the blue region in the past, consisting of events that never could have been affected by the traveler: it is the region outside the future light cones of all points on the traveler's world line.

These examples are a rather innocuous prelude to the mind-boggling physics of curved spacetime geometry, which also includes black holes. It took Einstein about ten years till 1915 to complete this second masterpiece, general relativity – one of the greatest intellectual achievements in the history of physics. It is surprising, to say the least, that although Einstein did receive a Nobel prize (in 1921), it was not for his theories of relativity. Such is the relativity of Nobel prizes! However, even without it Einstein stands out as one of the brightest, most creative and most fearless scientists ever.

Epilogue

The significant problems we have cannot be solved at the same level of thinking with which we created them.

Our visual journey through the landscape of space and time, which offered us a close view of the young Einstein's revolutionary insights, has come to an end. Isn't it remarkable that elementary reasoning, carefully carried through, can lead to such a counterintuitive, radically novel interpretation of physical reality? We have emphasized a geometric rather than an algebraic approach throughout, by consistently framing our explanations in the pictorial language of spacetime diagrams. This allowed us to tackle a number of famous paradoxes and to arrive at their sometimes surprising resolutions. If this method has allowed you to share some of the deeper views on nature relativity has to offer, then it has served its purpose. It is not accidental that the geometric approach works so well, because the theories of relativity drove a large part of physics into the arena of geometry, albeit a more general kind of geometry then the ancients dreamed of.

The roots of special relativity point through the work of Lorentz to Maxwell's theory of electromagnetism, which was completed around 1865. You may have wondered what happens to electromagnetic phenomena in relativity. Maxwell's theory is by construction already Lorentz invariant, and so are the descriptions it provides of electromagnetic phenomena. Still, it would be interesting to consider for example what electric and magnetic fields look like for different observers. I have refrained from discussing this subject because it requires a fair amount of knowledge of electromagnetism to start with.

After completing his theory of special relativity, Einstein moved on to other scientific questions. As we mentioned before, only later he returned to the problem of relativity, culminating in his general theory of relativity, which was completed in 1915. In this theory the notion of a flat spacetime, with which we were concerned in this book, was further generalized to curved spacetimes. The equivalence of inertial frames was extended to arbitrary frames, and the invariance under Lorentz transformations extended to invariance under general coordinate transformations. This mighty theory also introduced a radically new interpretation of the force of gravity as a manifestation of spacetime curvature. It suggested the existence of a number of astounding new physical effects and phenomena, which in the meantime have been vindicated by experiment. The most dramatic predictions are probably the existence of an expanding universe, black holes, and a cosmological constant or non-vanishing vacuum energy which permeates all of space.

The theory of relativity underscores the importance of the realization that our perceptions are by no means absolute. During our journey we have had to give up absolute notions of time, space, mass and energy. Yet this dependence of perspective on the observer is far from arbitrary; it is not the type of subjectivity we refer to when we say something is "a matter of taste". The theory of relativity provides us with an intersubjective meta-perspective, which transcends the single observer's viewpoint to hold for the collection of all observers. In that way relativity expresses a deep sense of universality.

Not long after Einstein's breakthrough, the foundations of physics were deeply shaken once more with the advent of quantum theory. Surprisingly, in that drastic revision of Newtonian physics the role of observers and the act of measurement lost even more of their objective meaning, to the extent that the strict separation between object and subject, which was still applicable in Einstein's theory, had to be given up. The new cornerstones of modern physics – relativity and quantum theory – had a deep and lasting impact on the philosophy of science and knowledge. They represent crucial turning points in our thinking that could not have been guessed from general considerations or philosophical inquiry: one had to go and study the actual physics itself in detail, like those great men did in the first quarter of the twentieth century.

About the scientific search for these fundamental laws, Einstein said: "There is no logical path that leads to these elementary laws, only an intuitive one, based on creativity and experience." He also observed cannily that "with such a methodological uncertainty, one would think that an arbitrary number of equally valid systems would be possible. However, history shows that of all conceivable constructions, always one stood out as absolutely superior to all others."

Retracing Einstein's steps in our very special way has hopefully given you a flavor of how exciting and rewarding it can be to be the first person to enter an unfamiliar territory of deep knowledge. Even today, vast but hidden domains of nature remain to be discovered. I can only hope that in generations to come, many will have the inspiration and courage to further explore the heart of this great domain full of secrets we call Nature.

Literature

The original papers by Einstein on special relativity are:

- Einstein, A., 'Zur Elektrodynamik bewegter Körper', *Annalen der Physik* 17, pp. 891-921, 1905.
- Einstein, A., 'Ist die Trägheit eines Körpers von seinem Energieinhalt abhängig?', *Annalen der Physik* 18, pp. 639-641, 1905.

English translations of all his 1905 papers with an introduction may be found in:

- Stachel, J. (ed.), *Einstein's miraculous year: five papers that changed the face of physics*, Princeton University Press, 2005.

Relatively accessible accounts of relativity by Einstein himself are:

- Einstein, A., *Relativity,* Prometheus Books, 1995 (1st edition 1916).
- Einstein, A., *The principle of relativity,* Dover, 1952.

Here is a selection of some of the many introductory books on special relativity:

- Bondi, H., *Relativity and Common Sense: A New Approach to Einstein*, Dover, 1980.
- French, A.P., *Special Relativity,* Norton, 1968.
- Mermin, N.D., *It's About Time, Understanding Einstein's relativity,* Princeton University Press, 2005.
- Møller, Christian, *The Theory of Relativity,* Clarendon Press, 1972 (original 1952).
- Resnick, R., *Introduction to Special Relativity,* Wiley, 1968.
- Synge, J.L., *Relativity: The Special Theory,* North-Holland, 1956.
- Taylor, E.F. & J.A. Wheeler, *Spacetime Physics – Introduction to Special Relativity,* Freeman, 1992.

Index

the most important page numbers are in **bold**

DISCARD